WEAK INTERACTIONS *and* MODERN PARTICLE THEORY

Howard Georgi
Mallinckrodt Professor of Physics
Harvard University

DOVER PUBLICATIONS, INC.
MINEOLA, NEW YORK

Bibliographical Note

This Dover edition, first published in 2009, is an unabridged republication of the work originally published in 1984 by the Benjamin/Cummings Publishing Company, Menlo Park, California.

Library of Congress Cataloging-in-Publication Data

Georgi, Howard.
Weak interactions and modern particle theory / Howard Georgi.—Dover ed.
p. cm.
Includes bibliographical references.
ISBN-13: 978-0-486-46904-1
ISBN-10: 0-486-46904-2
1. Weak interactions (Nuclear physics). I. Title.

QC794. 2009
539.7'544—dc22

2008042845

Manufactured in the United States by Courier Corporation
46904203
www.doverpublications.com

Contents

WEAK INTERACTIONS *and* MODERN PARTICLE THEORY

1 — Classical Symmetries

The concept of symmetry will play a crucial role in nearly all aspects of our discussion of weak interactions. At the level of the dynamics, the fundamental interactions (or at least that subset of the fundamental interactions that we understand) are associated with "gauge symmetries". But more than that, the underlying mathematical language of relativistic quantum mechanics — quantum field theory — is much easier to understand if you make use of all the symmetry information that is available. In this course, we will make extensive use of symmetry as a mathematical tool to help us understand the physics. In particular, we make use of the language of representations of Lie algebras.

1.1 Noether's Theorem – Classical

At the classical level, symmetries of an action which is an integral of a local Lagrangian density are associated with conserved currents. Consider a set of fields, $\phi_j(x)$ where $j = 1$ to N, and an action

$$S[\phi] = \int \mathcal{L}(\phi(x), \partial_\mu \phi(x))\, d^4x \tag{1.1.1}$$

where $\mathcal{L}$ is the local Lagrangian density. The index, j, is what particle physicists call a "flavor" index. Different values of j label different types, or "flavors", of the field ϕ. Think of the field, ϕ, without any explicit index, as a column vector in flavor space. Assume, for simplicity, that the Lagrangian depends only on the fields, ϕ, and their first derivatives, $\partial_\mu \phi$. The equations of motion are

$$\partial_\mu \frac{\delta \mathcal{L}}{\delta(\partial_\mu \phi)} = \frac{\delta \mathcal{L}}{\delta \phi}\,. \tag{1.1.2}$$

Note that (1.1.2) is a vector equation in flavor space. Each side is a row vector, carrying the flavor index, j.

A symmetry of the action is some infinitesimal change in the fields, $\delta\phi$, such that

$$S[\phi + \delta\phi] = S[\phi]\,, \tag{1.1.3}$$

or

$$\mathcal{L}(\phi + \delta\phi, \partial_\mu \phi + \delta \partial_\mu \phi) = \mathcal{L}(\phi, \partial_\mu \phi) + \partial_\mu V^\mu(\phi, \partial_\mu \phi, \delta\phi)\,, \tag{1.1.4}$$

where V^μ is some vector function of the order of the infinitesimal, $\delta\phi$. We assume here that we can throw away surface terms in the d^4x integral so that the V^μ terms makes no contribution to the action. But

$$\mathcal{L}(\phi + \delta\phi, \partial_\mu \phi + \delta \partial_\mu \phi) - \mathcal{L}(\phi, \partial_\mu \phi) = \frac{\delta \mathcal{L}}{\delta \phi}\, \delta\phi + \frac{\delta \mathcal{L}}{\delta(\partial_\mu \phi)}\, \partial_\mu \delta\phi\,, \tag{1.1.5}$$

because $\delta\partial_\mu\phi = \partial_\mu\delta\phi$. Note that (1.1.5) is a single equation with no j index. The terms on the right hand side involve a matrix multiplication in flavor space of a row vector on the left with a column vector on the right. From (1.1.2), (1.1.4) and (1.1.5), we have

$$\partial_\mu N^\mu = 0 \,, \tag{1.1.6}$$

where

$$N^\mu = \frac{\delta \mathcal{L}}{\delta(\partial_\mu \phi)}\, \delta\phi - V^\mu \,. \tag{1.1.7}$$

Often, we will be interested in symmetries that are symmetries of the Lagrangian, not just the action, in which case $V^\mu = 0$. In particular, our favorite symmetry will be linear unitary transformations on the fields, for which

$$\delta\phi = i\epsilon_a T^a \phi \,, \tag{1.1.8}$$

where the T^a for $a = 1$ to m are a set of $N \times N$ hermitian matrices acting on the flavor space, and the ϵ_a are a set of infinitesimal parameters. We can (and sometimes will) exponentiate (1.1.8) to get a finite transformation:

$$\phi \to \phi' = e^{i\epsilon_a T^a} \phi \,, \tag{1.1.9}$$

which reduces to (1.1.8) for small ϵ_a. The T^a's are then said to be the "generators" of the transformations, (1.1.9)

We will be using the T^a's so much that it is worth pausing to consider their properties systematically. The fundamental property of the generators is their commutation relation,

$$[T^a, T^b] = i\, f_{abc}\, T^c \,, \tag{1.1.10}$$

where f_{abc} are the structure constants of the Lie algebra, defined in any nontrivial representation (a trivial representation is one in which $T^a = 0$, for which (1.1.10) is trivially satisfied). The generators can then be classified into sets, called simple subalgebras, that have nonzero commutators among themselves, but that commute with everything else. For example, there may be three generators, T^a for $a = 1$ to 3 with the commutation relations of $SU(2)$,

$$[T^a, T^b] = i\, \epsilon_{abc}\, T^c \,, \tag{1.1.11}$$

and which commute with all the other generators. Then this is an $SU(2)$ factor of the algebra. The algebra can always be decomposed into factors like this, called "simple" subalgebras, and a set of generators which commute with everything, called $U(1)$'s.

The normalization of the $U(1)$ generators must be set by some arbitrary convention. However, the normalization of the generators of each simple subgroup is related to the normalization of the structure constants. It is important to normalize them so that in each simple subalgebra,

$$\sum_{c,d} f_{acd}\, f_{bcd} = k\, \delta_{ab} \,. \tag{1.1.12}$$

Then for every representation,

$$\operatorname{tr}(T^a T^b) \propto \delta_{ab} \,. \tag{1.1.13}$$

The mathematician's convention is to grant a special status to the structure constants and choose $k = 1$ in (1.1.12). However, we physicists are more flexible (or less systematic). For $SU(n)$, for example, we usually choose k so that

$$\operatorname{tr}(T^a T^b) = \frac{1}{2}\, \delta_{ab} \,. \tag{1.1.14}$$

for the n dimensional representation. Then $k = n$.

A symmetry of the form of (1.1.8) or (1.1.9) which acts only on the flavor space and not on the space-time dependence of the fields is called an "internal" symmetry. The familiar Poincare symmetry of relativistic actions is **not** an internal symmetry.

In this book we will distinguish two kinds of internal symmetry. If the parameters, ϵ_a, are independent of space and time, the symmetry is called a "global" symmetry. Global symmetries involve the rotation of ϕ in flavor space in the same way at all points in space and at all times. This is not a very physically appealing idea, because it is hard to imagine doing it, but we will see that the concept of global symmetry is enormously useful in organizing our knowledge of field theory and physics.

Later on, we will study what happens if the ϵ_a depend on x. Then the symmetry is a "local" or "gauge" symmetry. As we will see, a local symmetry is not just an organizing principle, but is intimately related to dynamics.

For now, consider (1.1.7) for a global internal symmetry of $\mathcal{L}$. Because the symmetry is a symmetry of $\mathcal{L}$, the second term, V^μ is zero.

Thus we can write the conserved current as

$$N^\mu = \frac{\delta \mathcal{L}}{\delta(\partial_\mu \phi)}\, i\epsilon_a T^a \phi \,. \tag{1.1.15}$$

Because the infinitesimal parameters are arbitrary, (1.1.15) actually defines m conserved currents,

$$J_a^\mu = -i\frac{\delta \mathcal{L}}{\delta(\partial_\mu \phi)}\, T^a \phi \quad \text{for } a = 1 \text{ to } m. \tag{1.1.16}$$

I stress again that this discussion is all at the level of the classical action and Lagrangian. We will discuss later what happens in quantum field theory.

1.2 Examples

Example 1

Let Φ_j, for $j = 1$ to N, be a set of real scalar boson fields. The most general possible real quadratic term in the derivatives of Φ in the Lagrangian is

$$\mathcal{L}_{KE}(\Phi) = \frac{1}{2}\, \partial_\mu \Phi^T\, S\, \partial^\mu \Phi \,. \tag{1.2.1}$$

where S is a real symmetric matrix. Note the matrix notation, in which, the Φ_j are arranged into an N-component column vector. In physical applications, we want S to be a strictly positive matrix. Negative eigenvalues would give rise to a Hamiltonian that is not bounded below and zero eigenvalues to scalars that do not propagate at all. If S is positive, then we can define a new set of fields by a linear transformation

$$L\,\phi \equiv \Phi \,, \tag{1.2.2}$$

such that

$$L^T\, S\, L = I \,. \tag{1.2.3}$$

In terms of ϕ, the Lagrangian becomes

$$\mathcal{L}_{KE}(\phi) = \frac{1}{2}\, \partial_\mu \phi^T\, \partial^\mu \phi \,. \tag{1.2.4}$$

This is the canonical form for the Lagrangian for a set of N massless free scalar fields. Under an infinitesimal linear transformation,

$$\delta\phi = G\,\phi\,, \tag{1.2.5}$$

where G is an $N \times N$ matrix, the change in $\mathcal{L}_{KE}$ is

$$\delta\mathcal{L}_{KE}(\phi) = \frac{1}{2}\,\partial_\mu\phi^T\,(G+G^T)\,\partial^\mu\phi\,. \tag{1.2.6}$$

If G is antisymmetric, the Lagrangian is unchanged! We must also choose G real to preserve the reality of the fields. Thus the Lagrangian is invariant under a group of $SO(N)$ transformations,

$$\delta\phi = i\epsilon_a T^a \phi\,, \tag{1.2.7}$$

where the T^a, for $a = 1$ to $N(N-1)/2$ are the $N(N-1)/2$ independent, antisymmmetric imaginary matrices. These matrices are a representation of the $SO(N)$ algebra.[1] When exponentiated, (1.2.7) produces an orthogonal transformation, a representation of the group $SO(N)$

$$\phi \to \phi' = O\,\phi\,, \tag{1.2.8}$$

where

$$O^T = O^{-1}\,. \tag{1.2.9}$$

This is a rotation in a real N-dimensional space.

Notice that we have not had to do anything to impose this $SO(N)$ symmetry except to put the Lagrangian into canonical form. It was an automatic consequence of the physical starting point, the existence of N free massless scalar fields. The canonical form of the derivative term in the kinetic energy automatically has the $SO(N)$ symmetry.

The corresponding Noether currents are

$$J^\mu_a = -i(\partial^\mu\phi^T)T^a\phi\,. \tag{1.2.10}$$

The symmetry, (1.2.8), is the largest internal rotation symmetry that a set of N real spinless bosons can have, because the kinetic energy term, (1.2.4), must always be there. However, the symmetry may be broken down to some subgroup of $SO(N)$. This can happen trivially because of the mass term if the scalars are not all degenerate. The mass term has the general form:

$$\mathcal{L}_{mass}(\phi) = -\frac{1}{2}\,\phi^T M^2\,\phi\,, \tag{1.2.11}$$

where M^2 is a real symmetric matrix called the mass matrix. Its eigenvalues (if all are positive) are the squared masses of the scalar particles. The mass term is invariant under an orthogonal transformation, O, if

$$O^T M^2 O = M^2 \quad \text{or equivalently} \quad [O, M^2] = 0\,. \tag{1.2.12}$$

If M^2 is proportional to the identity matrix, then the entire $SO(N)$ is unbroken. In general, the transformations satisfying (1.2.12) form a representation of some subgroup of $SO(N)$. The subgroup is generated by the subset of the generators, T^a, which commute with M^2.

[1]Usually, I will not be careful to distinguish an algebra or group from its representation, because it is almost always the explicit representation that we care about.

The mass matrix can be diagonalized by an orthogonal transformation that leaves the derivative term, (1.2.4), unchanged. Then the remaining symmetry is an $SO(\ell)$ for each ℓ degenerate eigenvalues. For example, if K of the fields have mass m_1 and the other $N-K$ have mass m_2, the diagonal mass matrix can be taken to have to form

$$M^2 = \begin{pmatrix} m_1^2 & & & & & \\ & \ddots & & & 0 & \\ & & m_1^2 & & & \\ & & & m_2^2 & & \\ & 0 & & & \ddots & \\ & & & & & m_2^2 \end{pmatrix} . \tag{1.2.13}$$

The symmetry is an $SO(K) \times SO(N-K)$ which rotates the degenerate subsets, of the form

$$O = \begin{pmatrix} O_1 & 0 \\ 0 & O_2 \end{pmatrix} , \tag{1.2.14}$$

where O_1 and O_2 act on the fields with mass m_1 and m_2 respectively. The finite group elements, (1.2.14), are generated by exponentiation of the $K(K-1)/2 + (N-K)(N-K-1)/2$ generators

$$\begin{pmatrix} S_1 & 0 \\ 0 & 0 \end{pmatrix} \quad \text{and} \quad \begin{pmatrix} 0 & 0 \\ 0 & S_2 \end{pmatrix} , \tag{1.2.15}$$

where S_1 and S_2 are antisymmetric imaginary matrices.

If the mass matrix is non-degenerate, that is with no pair of eigenvalues equal, the $SO(N)$ symmetry is completely broken and there is no continuous symmetry that remains, although the Lagrangian is still invariant under the discrete symmetry under which any component of ϕ changes sign.

The symmetry can also be broken down by interaction terms in the Lagrangian. With cubic and quartic terms in ϕ, the symmetry can be broken in more interesting ways. Consider, as a simple example that will be of use later, a Lagrangian with $N = 8$ scalars, with $K = 4$ with mass m_1 (the ϕ_j for $j = 1$ to 4) and the rest with mass m_2 (the ϕ_j for $j = 5$ to 8). The kinetic energy and mass terms then have the symmetry, $SO(4) \times SO(4)$. Arrange the 8 real fields into two complex doublet fields as follows:

$$\xi_1 \equiv \begin{pmatrix} (\phi_1 + i\phi_2)/\sqrt{2} \\ (\phi_3 + i\phi_4)/\sqrt{2} \end{pmatrix} , \quad \xi_2 \equiv \begin{pmatrix} (\phi_5 + i\phi_6)/\sqrt{2} \\ (\phi_7 + i\phi_8)/\sqrt{2} \end{pmatrix} . \tag{1.2.16}$$

Now consider an interaction term of the form

$$\mathcal{L}_{int} = \lambda \, \xi_1^\dagger \xi_2 \, \xi_2^\dagger \xi_1 . \tag{1.2.17}$$

This interaction term is not invariant under the $SO(4) \times SO(4)$, but it is invariant under an $SU(2) \times U(1)$ symmetry under which

$$\xi_j \to U \, \xi_j \quad \text{for } j = 1 \text{ to } 2, \tag{1.2.18}$$

where the 2×2 matrix, U, is unitary. The U can written as a phase (the $U(1)$ part) times a special unitary matrix V ($\det V = 1$), so it is a representation of $SU(2) \times U(1)$. This is the subgroup of $SO(4) \times SO(4)$ left invariant by the interaction term, (1.2.17).

The symmetry structure of a field theory is a descending hierarchy of symmetry, from the kinetic energy term down through the interaction terms. The kinetic energy term has the largest possible symmetry. This is broken down to some subgroup by the mass term and the interaction terms. This way of looking at the symmetry structure is particularly useful when some of the interactions terms are weak, so that their effects can be treated in perturbation theory.

Example 2

Let ψ_j for $j = 1$ to N be free, massless, spin-$\frac{1}{2}$, four-component Dirac fermion fields with Lagrangian

$$\mathcal{L}(\psi) = i\,\overline{\psi}\,\partial\!\!\!/\,\psi\,. \tag{1.2.19}$$

The Dirac fermion fields are complex, thus (1.2.19) has an obvious $SU(N) \times U(1)$ symmetry under which, for infinitesimal changes,

$$\delta\psi = i\epsilon_a T^a \psi\,, \tag{1.2.20}$$

where the T^a are the N^2 hermetian $N \times N$ matrices, which generate the defining (or N) representation of $SU(N) \times U(1)$. When exponentiated this gives

$$\psi \to U\,\psi\,, \tag{1.2.21}$$

where U is a unitary $N \times N$ matrix

$$U = e^{i\epsilon_a T^a}\,. \tag{1.2.22}$$

The $SU(N)$ part of U is generated by the $N^2 - 1$ traceless hermetian matrices, while the generator of the $U(1)$ part, which commutes with everything, is proportional to the identity. The $U(1)$ is just fermion number.

In (1.2.19),

$$\frac{\delta\mathcal{L}}{\delta(\partial_\mu\psi)} = i\overline{\psi}\gamma^\mu\,, \tag{1.2.23}$$

hence the Noether current is

$$J_a^\mu = \overline{\psi}\gamma^\mu T^a \psi\,. \tag{1.2.24}$$

With $T^a = 1$, the conserved Noether current associated with the $U(1)$ is just the fermion number current.

The transformation, (1.2.21), is by no means the largest internal symmetry of (1.2.19). To see the rest of the symmetry, and for many other reasons later on, we will make extensive use of two-component, Weyl fermion fields, which are related to ordinary spin-$\frac{1}{2}$, four-component fermion fields, ψ, by projection with the projection operators

$$P_\pm \equiv (1 \pm \gamma_5)/2\,. \tag{1.2.25}$$

Define the left-handed (L) and right-handed (R) fields as

$$\psi_L \equiv P_+\psi\,, \quad \psi_R \equiv P_-\psi\,. \tag{1.2.26}$$

Then

$$\overline{\psi_L} = \overline{\psi}P_-\,, \quad \overline{\psi_R} = \overline{\psi}P_+\,. \tag{1.2.27}$$

But then because

$$P_+\gamma^\mu = \gamma^\mu P_-\,, \tag{1.2.28}$$

we can write

$$\mathcal{L}(\psi) = i\,\overline{\psi}\,\partial\!\!\!/\,\psi = i\,\overline{\psi_L}\,\partial\!\!\!/\,\psi_L + i\,\overline{\psi_R}\,\partial\!\!\!/\,\psi_R\,. \tag{1.2.29}$$

Evidently, we have the freedom to make separate $SU(N) \times U(1)$ transformations on the L and R fermions:

$$\delta\psi_L = i\epsilon_a^L T^a \psi_L\,, \quad \delta\psi_R = i\epsilon_a^R T^a \psi_R\,, \tag{1.2.30}$$

or

$$\psi_L \to U_L \psi_L \,, \quad \psi_R \to U_R \psi_R \,. \tag{1.2.31}$$

The physical interpretation of the L and R states is that they are helicity eigenstates for on-mass-shell fermions. Consider a plane wave moving in the 3 direction. The fermions are massless, thus $p^0 = p^3$ while $p^1 = p^2 = 0$. The Dirac equation in momentum space is $\not{p}\psi = 0$, or $p^0(\gamma^0 - \gamma^3)\psi = 0$, or

$$\gamma^0\psi = \gamma^3\psi \,. \tag{1.2.32}$$

The spin angular momentum in the 3 direction is

$$J^3 = \frac{\sigma^{12}}{2} = \frac{i\gamma^1\gamma^2}{2} \,. \tag{1.2.33}$$

Then

$$\begin{aligned} J^3\psi_L &= \tfrac{i}{2}\,\gamma^1\gamma^2\psi_L = \tfrac{i}{2}\,\gamma^0\gamma^0\gamma^1\gamma^2\psi_L \\ &= \tfrac{i}{2}\,\gamma^0\gamma^1\gamma^2\gamma^0\psi_L = \tfrac{i}{2}\,\gamma^0\gamma^1\gamma^2\gamma^3\psi_L \\ &= -\tfrac{1}{2}\,\gamma_5\psi_L = -\tfrac{1}{2}\psi_L \,. \end{aligned} \tag{1.2.34}$$

Thus ψ_L describes a particle with helicity $-\frac{1}{2}$, that is a left handed particle, while ψ_R describes a particle with helicity $\frac{1}{2}$, that is a right handed particle.

The symmetry, (1.2.30), under which the L and R fields transform differently, is called a **chiral symmetry**. Note that the generators of the chiral symmetry all commute with the fermion number.

The chiral symmetry is still not the largest internal symmetry of (1.2.19). To identify the largest symmetry, we must introduce the idea of charge conjugate fields. Define

$$\psi_c = C\psi^* \,, , \tag{1.2.35}$$

and

$$\psi_{cL} = P_+\psi_c \,, \quad \psi_{cR} = P_-\psi_c \,, \tag{1.2.36}$$

where C is the charge conjugation matrix acting on the Dirac indices, satisfying

$$C^2 = 1 \,, \quad C^\dagger = C \,, \quad C\,\gamma^{\mu *}\,C = -\gamma^\mu \,. \tag{1.2.37}$$

We will sometimes take C to be nontrivial in flavor space as well, but for now, we will assume that it acts in the space of the Dirac indices. The form of C depends on the representation of the γ matrices. For example, in the so-called Majorana representation in which all the γ's are imaginary, $C = 1$, while in the standard representation of Bjorken and Drell, $C = i\gamma^2$.

The name "charge conjugation" comes originally from the effect of the transformation (1.2.35) on the Dirac equation for a particle in an electromagnetic field. Such a particle, with mass m and charge e, satisfies

$$i\not{\partial}\psi - e\not{A}\psi - m\psi = 0 \,, \tag{1.2.38}$$

where A^μ is the (hermetian) electromagnetic field. Taking the complex conjugate, multiplying by C and using (1.2.35) gives

$$i\not{\partial}\psi_c + e\not{A}\psi_c - m\psi_c = 0 \,. \tag{1.2.39}$$

The C operation changes the particle of charge e to its antiparticle, with the same mass but charge $-e$.

The interesting thing about charge conjugation from our present point of view is that it changes L to R,

$$\begin{aligned} C\psi_R^* &= C\,(P_-)^*\,\psi^* = C\left(\tfrac{1-\gamma_5^*}{2}\right)\psi^* \\ &= C\left(\tfrac{1-\gamma_5^*}{2}\right)C\,C\,\psi^* = P_+\psi_c = \psi_{cL}\,, \end{aligned} \tag{1.2.40}$$

where we have used $C\gamma_5^*C = -\gamma_5$. This is what we need to find a larger symmetry of (1.2.19). The point is that with a purely internal symmetry, we cannot mix up the ψ_L with the ψ_R. A transformation such as

$$\delta\psi_L = i\epsilon_a T^a \psi_R \tag{1.2.41}$$

makes no sense at all. You can see this trivially by acting on both sides with P_-.

But if we first change the ψ_R into ψ_{cL} with charge conjugation, then we can mix them with the ψ_L. Such a transformation will not commute with the $U(1)$ of fermion number, because ψ and ψ_c have opposite fermion number, but at this point, we do not care.

We can use (1.2.40) to show that

$$\mathcal{L}(\psi) = i\,\overline{\psi_L}\,\partial\!\!\!/\,\psi_L + i\,\overline{\psi_{cL}}\,\partial\!\!\!/\,\psi_{cL} + \text{total derivative.} \tag{1.2.42}$$

We drop the total derivative, which does not effect the action, and then combine the L fields into a $2N$ component column vector:

$$\Psi_L = \begin{pmatrix} \psi_L \\ \psi_{cL} \end{pmatrix}. \tag{1.2.43}$$

In terms of Ψ, the Lagrangian is

$$\mathcal{L}(\Psi) = i\,\overline{\Psi}\,\partial\!\!\!/\,\Psi\,. \tag{1.2.44}$$

Clearly, the Lagrangian, (1.2.44), in invariant under $SU(2N)\times U(1)$ transformations, generated by hermetian $2N\times 2N$ matrices acting on the extended flavor space of (1.2.43). Note that if fermion number is not conserved, there is not even any reason that there should be an even number of L fields!. For example, in the standard model, we seem to have 3 approximately massless left-handed neutrinos, each described by a two-component Weyl field.

Most of the Lagrangians that we actually deal with will have interactions that distinguish between particles and their antiparticles, thus breaking the symmetries that mix fermions with antifermions. For example, if the fermions have a charge, as in QED, the symmetry is broken down to a chiral symmetry, (1.2.30), because the fact that the charge changes sign under charge conjugation breaks the larger symmetry.

Fermion Masses

The most general fermion mass term that preserves fermion number has the form,

$$\mathcal{L}_{mass} = -\overline{\psi_L}\,M\,\psi_R + \overline{\psi_R}\,M^\dagger\,\psi_L\,, \tag{1.2.45}$$

where

$$\overline{\psi_L} = (\psi_L)^\dagger\gamma^0\,, \quad \overline{\psi_R} = (\psi_R)^\dagger\gamma^0\,, \tag{1.2.46}$$

and the mass matrix, M, is an arbitrary complex matrix. Note that the two terms in (1.2.45) are hermetian conjugates of one another. Because the ψ_L and ψ_R terms are coupled together, the Lagrangian is not invariant under the chiral symmetry, (1.2.30). A chiral transformation changes M

$$M \to U_L^\dagger M U_R\,. \tag{1.2.47}$$

However, that means that we can use the chiral symmetry to put the mass matrix, M, into a canonical form. We can diagonalize it and make each of the eigenvalues real and positive![2]

The symmetry of the mass term that commutes with fermion number is then determined by the degeneracy of the diagonal mass matrix. If all the masses are different, then the only symmetry that remains is a $U(1)^N$ symmetry, separate conservation of each of the fermion flavors. If ℓ flavors have the same mass, there is a non-Abelian $SU(\ell)$ symmetry. For example, if all N fermions are degenerate, the mass matrix is proportional to

$$\mathcal{L}_{mass} = \overline{\psi}\psi\,, \tag{1.2.48}$$

This breaks the chiral symmetry, (1.2.30), but it is still invariant under the $SU(N)\times U(1)$ symmetry, (1.2.20), in which the L and R fields rotate together.

If fermion number is not conserved, so that the fermions are most appropriately described by N left-handed Weyl fields, Ψ_L, the most general mass term has the form

$$\overline{\Psi_{cR}}\, M \Psi_L\,, \tag{1.2.49}$$

where M is a complex, symmetric matrix. The kinetic energy term is invariant under an $SU(N)\times U(1)$ symmetry:

$$\Psi_L \to U\Psi_L\,. \tag{1.2.50}$$

Under (1.2.50), the mass matrix, M changes to

$$M \to M' = U^T\, M\, U\,, \tag{1.2.51}$$

This can be used to make M diagonal and positive.

Problems

1.1. The Lagrangian for a set of N massless free scalar fields, (1.2.4), has a rather peculiar symmetry under which

$$\phi \to \phi + a\,,$$

for any constant a. What is the Noether current? Note that this "symmetry" really is peculiar. While it is a symmetry of the classical Lagrangian, it is not a symmetry of the corresponding quantum theory. This is one of the simplest examples of a continuous symmetry that is spontaneously broken. Spontaneous symmetry breaking is a subject to which we will return many times.

1-2. Find a basis for the γ^μ matrices in which the charge conjugation matrix, C, is the identity. This is called a Majorana representation. In this representation, we can define real 4-component fields

$$\psi_1 = (\psi + \psi_c)/\sqrt{2} = (\psi + \psi^*)/\sqrt{2}\,,$$

$$\psi_2 = i\,(\psi - \psi^*)/\sqrt{2}\,.$$

Rewrite the mass term for N degenerate Dirac fermions, (1.2.48), in terms of these real fields, and identify the symmetry of this term. What's the symmetry group? Show that this is also a symmetry of the kinetic energy term for free Dirac fermions.

1-3. Show that C in (1.2.37) is essentially unique in any given representation of the γ matrices.

[2] This is all at the classical level — we will come back to the question of real masses later on when we discuss the strong CP puzzle.

1a — Quantum Field Theory

Many of the results of the previous chapter can be carried over immediately to quantum field theory. However, the necessity of regularization and renormalization sometimes makes things more interesting. In this chapter, we discuss an approach to quantum field theory that is particularly useful for the discussion of weak interactions and the Standard model in general. We will assume that the reader is familiar with quantum field theory, and has available an encyclopedic text such as [Itzykson]. Here we will not pretend to anything like completeness, but will concentrate instead on explaining what is really going on. The detail of our regularization and renormalization scheme will seldom matter, but when it does, we will use dimensional regularization and minimal subtraction. A brief review will be included in an appendix to this book.

1a.1 Local Quantum Field Theory

Local field theory is a useful idealization. The only way we know to describe the quantum mechanical interactions of a finite number of types of particles in ordinary space-time is in a local quantum field theory characterized by a local Lagrangian density, with the interactions described by products of fields at the same space-time point.[1]

We strongly suspect that this picture is only approximate. It would be astonishing if our naive picture of the space-time continuum, or quantum mechanics, or anything else that the human mind can imagine, continued to make sense down to arbitrarily small distances. However, relativistic quantum mechanics appears to be an excellent approximation at the distance we can probe today. It is hard to see how deviations from the rules of relativistic quantum mechanics could have failed to show up in any of the myriad of experiments at high energy accelerators around the world, unless they are hidden because they are associated with physics at very tiny distances. Thus we are justified in adopting local quantum field theory as a provisional description of physics at the distances we can probe for the foreseeable future.

The fact that unknown physics may be lurking at very small distances is related to the necessity for regularization and renormalization of quantum field theory. A local Lagrangian in a quantum field theory is not *a priori* well-defined precisely because interactions that involve products of fields at a single space-time point specify the physics down to arbitrarily small distances. In mathematical language, the fields are distributions rather than functions, and multiplying them together at the same space-time point is a dangerous act. In regularization and renormalization, the physics at small distances is modified in some way to make the theory well-defined. Then the dependence on the short distance physics is incorporated into a set of parameters that can be related to physical quantities at measurable distances. A renormalizable theory is one in which only a finite number

[1] String theory may be a consistent description of the relativistic quantum mechanics of an infinite number of types of particles. However, because in a typical string theory, only a finite number of particles have masses small compared to the Planck scale, we can describe the low energy physics by a field theory.

of parameters are required to absorb all the dependence on short distance physics.

We will have much more to say about these issues as we develop the idea of effective field theory. Eventually, we will make the idea of hiding unknown physics at small distances into one of our principle tools. For now, we simply discuss renormalizable, local quantum field theory as if it were a God-given truth.

Let us begin by discussing a simple massless scalar field theory, defined by

$$\mathcal{L}(\phi) = \frac{1}{2}\,\partial^\mu\phi\partial_\mu\phi - \frac{\lambda}{4!}\,\phi^4 + s_\phi\phi\,. \tag{1a.1.1}$$

In addition to the quantum field, ϕ, with mass dimension 1, we have included a c-number source, s_ϕ, with mass dimension 3 for the scalar field ϕ. This is convenient, because it gives us a handle with which to tweak the theory to see how it responds. All the Green functions are implicitly present in the structure of the "vacuum" as a function of the source,

$$Z(s) = e^{iW(s)} = \langle 0 \mid 0\rangle_s\,. \tag{1a.1.2}$$

$W(s)$ is the generating function for the connected Green functions. Actually, as we will discuss later, it is not so easy to define this so-called $\lambda\phi^4$ theory beyond perturbation theory, because the theory is sick at short distances, but for now, we will be happy with a perturbative definition.

Without the source term, the action derived from (1a.1.1) is classically scale invariant. This means that it is invariant under the following transformation:

$$\phi(x) \to \phi'(x') \tag{1a.1.3}$$

where

$$x' = e^\lambda\,x \qquad\qquad \phi'(x') = e^{-\lambda}\,\phi(x) \tag{1a.1.4}$$

The classical scale invariance would be broken if we put in a mass term. We will leave the mass term out so that we have only one parameter to talk about — the coupling λ.

The first interesting feature that we encounter when we look at a quantum field theory defined by dimensional regularization and minimal subtraction (DRMS) is that scale invariance is broken. All the parameters in the theory depend on the renormalization scale, μ, used to define their dimensional extensions. The appearance of the parameter μ has an important physical consequence, which is easy to understand in perturbation theory. Quantities calculated in DRMS depend on $\ln\mu$. If all the other relevant dimensional parameters in the calculation are of the same order as μ, these logs cause no problems. However, if some dimensional parameters are very different from μ, perturbation theory will break down before it should. Thus calculations should be organized so that, as much as possible, all relevant dimensional parameters are of order μ. Fortunately, μ can be chosen at our convenience, and given the parameters for one value of μ, we can calculate the set of parameters for a different value of μ that lead to the same physics.

The Renormalization Group

The process of changing μ while keeping the physics unchanged is associated with the name "the renormalization group". This is unfortunate, because it makes it sound as if something nontrivial is going on, when in fact the process is quite simple. The only "group" structure involved is the group of one dimensional translations. The important (and extremely trivial) fact is that a large translation can be built up by putting small ones together! Consider the parameter λ in (1a.1.1). If we know λ at μ_1, then we can compute it at μ_2 using perturbation theory:

$$\lambda(\mu_2) = L(\lambda(\mu_1), \ln(\mu_2/\mu_1))\,, \tag{1a.1.5}$$

for some function L. If both $\lambda(\mu_1)$ and $\ln(\mu_2/\mu_1)$ are small, we can reliably compute L in perturbation theory. However, so long as both $\lambda(\mu_1)$ and $\lambda(\mu_2)$ are small, we can dispense with the restriction that $\ln(\mu_2/\mu_1)$ is small. Instead of using (1a.1.5) directly, we can get from μ_1 to μ_2 in a series of small steps. The lowest order perturbative result is

$$\lambda(\mu_2) = \lambda(\mu_1) + B\,\lambda(\mu_1)^2 \ln\left(\frac{\mu_2}{\mu_1}\right) + \mathcal{O}(\lambda^3), \tag{1a.1.6}$$

for some constant B.

Let

$$\Delta = \Delta(\ln(\mu)) \equiv \frac{1}{N} \ln\left(\frac{\mu_2}{\mu_1}\right). \tag{1a.1.7}$$

Then

$$\begin{aligned}
\lambda(\mu_1 e^{\Delta}) &= \lambda(\mu_1) + B\,\lambda(\mu_1)^2\,\Delta + \mathcal{O}(\lambda^3), \\
\lambda(\mu_1 e^{2\Delta}) &= \lambda(\mu_1 e^{\Delta}) + B\,\lambda(\mu_1 e^{\Delta})^2\,\Delta + \mathcal{O}(\lambda^3), \\
\lambda(\mu_1 e^{3\Delta}) &= \lambda(\mu_1 e^{2\Delta}) + B\,\lambda(\mu_1 e^{2\Delta})^2\,\Delta + \mathcal{O}(\lambda^3), \\
&\cdots
\end{aligned} \tag{1a.1.8}$$

Putting these all together, we get

$$\lambda(\mu_2) = \lambda(\mu_1) + \sum_{j=0}^{N-1} B\,\lambda(\mu_1 e^{j\Delta})^2\,\Delta + \mathcal{O}(\lambda^3), \tag{1a.1.9}$$

or in the limit as $N \to \infty$,

$$\lambda(\mu_2) = \lambda(\mu_1) + \int_{\mu_1}^{\mu_2} B\,\lambda(\mu)^2 \frac{d\mu}{\mu} + \mathcal{O}(\lambda^3), \tag{1a.1.10}$$

We have arrived, by a rather backwards route, at the lowest order integral equation for the "running" coupling, $\lambda(\mu)$. Note that we have not done anything at all except to build up a large change in μ out of infinitesimal changes. If we differentiate with respect to μ_2 and set $\mu_2 = \mu$, we get the standard differential equation for the running coupling:

$$\mu \frac{\partial}{\partial \mu} \lambda(\mu) = \beta_\lambda(\lambda(\mu)), \tag{1a.1.11}$$

where

$$\beta_\lambda(\lambda) = B\,\lambda^2 + \mathcal{O}(\lambda^3) \tag{1a.1.12}$$

is the β-function. We will see below that $B > 0$, so that λ is an increasing function of μ for small λ.

Ignoring the terms of order λ^3, we can solve (1a.1.12) to obtain

$$\lambda(\mu) = \frac{\lambda(\mu_0)}{1 + B\,\lambda(\mu_0)\ln(\mu_0/\mu)} = \frac{1}{B\,\ln(\Lambda/\mu)}. \tag{1a.1.13}$$

This solution exhibits the most interesting feature of the running coupling, dimensional transmutation[Coleman 73]. Because the dimensionless quantity, λ, is a calculable function of the dimensional quantity, μ, the parameter that actually characterizes the physics is a dimensional one, Λ. (1a.1.13) also shows the difficulty with this theory. The scale μ cannot be taken arbitrarily large. Above $\mu \approx \Lambda$, the coupling is large and the theory is not defined perturbatively. There are good reasons to believe that this difficulty persists beyond perturbation theory and that the $\lambda\phi^4$ theory does not make sense at short distances.

This is sometimes discussed in terms of the rather ridiculous word, "triviality". The word is a holdover from an old fashioned way of looking at field theory. All it means is that the $\lambda\phi^4$ theory is not a useful description of the physics at arbitrarily short distances. If you try to define the theory in isolation, by cutting the physics off at some small distance, renormalizing, and then letting the distance go to zero, the theory is "trivial" because the renormalized coupling is driven to zero. This is simply the flip side of (1a.1.13) in which a finite renormalized coupling leads to a disaster at a finite small distance. In practice, this means that the physics has to change at some momentum scale smaller than the Λ in (1a.1.13). However, as we will see in more detail later on, this does not prevent us from making sense of the theory at scales at which $\lambda(\mu)$ is small.

The Renormalization Group Equation

We will begin by considering one particle irreducible (1PI) graphs. These are easy to calculate and useful. Later we will also discuss connected Green functions and their generating functional, $W(s)$. Call a 1PI function with n external ϕ lines, $\Gamma_n(p_i, \lambda, \mu)$, where we have indicated the dependence on the external momenta, p_i for $i = 1$ to n, on the parameter λ, and on the scale μ. If we change μ, we must change the wave function renormalization of each of the external lines and also the coupling to maintain the same physics. This is the conveniently summarized in the renormalization group equation:

$$\left(\mu\frac{\partial}{\partial\mu} + \beta_\lambda\frac{\partial}{\partial\lambda} - n\,\gamma_\phi\right)\Gamma_n = 0\,. \tag{1a.1.14}$$

In addition to β_λ, (1a.1.14) depends on an "anomalous dimension", γ_ϕ, for the ϕ field. Both of these are functions of λ.

Let us now actually calculate β_λ and γ_ϕ to one-loop order. The simplest way to do this, is to calculate the 1PI functions for small n, and demand that (1a.1.14) be satisfied.

To one loop, the Feynman graphs that contribute to the 2-point function, Γ_2, are

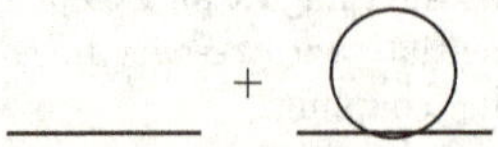

(1a.1.15)

This vanishes in DRMS for a massless scalar (and in any sensible scheme, it does not depend on the external momentum) and thus the wave function renormalization is 1 in one loop and $\gamma_\phi = 0$ to order λ.

The Feynman graphs that contribute the Γ_4 to one loop are

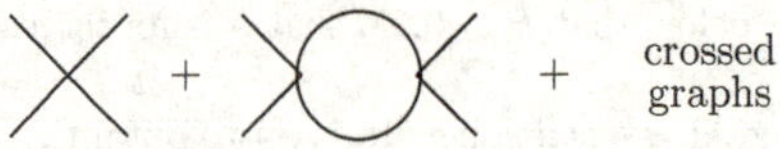

(1a.1.16)

The result is (with all momenta going in)

$$\begin{aligned}\Gamma_4(p_i, \lambda, \mu) = \lambda + \frac{\lambda^2}{32\pi^2}&\left[\ln\left(\frac{-(p_1+p_2)^2}{\mu^2}\right) + \ln\left(\frac{-(p_1+p_3)^2}{\mu^2}\right)\right.\\ &\left.+\ln\left(\frac{-(p_1+p_4)^2}{\mu^2}\right)\right] + \cdots\,,\end{aligned} \tag{1a.1.17}$$

where the $\cdots$ are non-logarithmic terms.

Now when (1a.1.17) is inserted into (1a.1.14), we find

$$\beta_\lambda - \frac{3\lambda^2}{16\pi^2} + \mathcal{O}(\lambda^3) = 0\,, \tag{1a.1.18}$$

or

$$\beta_\lambda = \frac{3\lambda^2}{16\pi^2} + \mathcal{O}(\lambda^3)\,. \tag{1a.1.19}$$

It is worth commenting on some minus signs that frequently cause confusion. Note that the μ dependence of $\lambda(\mu)$ is opposite to that of Γ_4. Because the wave function renormalization vanishes, Γ_4 is a directly physical scattering amplitude, to this order. Thus when μ changes, Γ_4 remains constant. That is a direct translation of (1a.1.14), with $\gamma_\phi = 0$.

On the other hand, the "solution", to the renormalization group equation, is

$$\Gamma_4(p_i, \lambda(\mu), \mu) = \Gamma_4(p_i, \lambda(\mu_0), \mu_0)\,, \tag{1a.1.20}$$

for $\lambda(\mu)$ satisfying (1a.1.13). Since $\lambda(\mu_0)$ can be regarded as a function of $\lambda(\mu)$ and μ, (1a.1.20) says that Γ_4 is not a function of $\lambda(\mu)$ and μ separately, but only of the combination, $\lambda(\mu_0)$. The μ dependence of λ is then related, in the absence of wave function renormalization, not to the μ dependence of Γ_4, but to its dependence on the scale of p, because it is when $p \approx \mu$, that $\Gamma_4 \approx \lambda(\mu)$.

Sources and Parameters

The form, (1a.1.14), obscures a connection between the anomalous dimension, γ_ϕ, and the function, β_λ. First of all, the sign in front of the anomalous dimension term depends on the fact we are studying 1PI functions, Γ_n, rather than connected Green functions, G_n, which satisfy

$$\left(\mu\frac{\partial}{\partial\mu} + \beta_\lambda\frac{\partial}{\partial\lambda} + n\,\gamma_\phi\right) G_n = 0\,. \tag{1a.1.21}$$

Secondly, because $W(s)$ is the generating function for the connected Green functions, we can obtain (1a.1.21) for all n from the single equation

$$\left(\mu\frac{\partial}{\partial\mu} + \beta_\lambda\frac{\partial}{\partial\lambda} + \int \beta_{s(x)}\frac{\partial}{\partial s(x)}\, d^4x\right) W(\lambda, s) = 0\,. \tag{1a.1.22}$$

We can get back to (1a.1.21) by functionally differentiating with respect to s. Thus

$$\beta_{s(x)} = s(x)\,\gamma_\phi\,. \tag{1a.1.23}$$

The General Case — Mixing

In general, there are many couplings, λ_i, many fields, ϕ_α, and corresponding sources, s_α. The nice thing about the form, (1a.1.22), is that it generalizes trivially to the general case. The parameters, λ_i, may include mass terms, terms of dimension 3, whatever is around. In fact, as we will see, this even makes sense in nonrenormalizable theories, properly defined. The β's depend on the λ parameters and the sources, s_α. In a mass independent renormalization scheme such as DRMS, there is no μ dependence in any of these functions, and thus the dependence on the parameters is strongly constrained by dimensional analysis. In perturbation theory, each β is a sum of products of λ's and s's (derivatives acting on the s's are also possible, although they do not appear in this simple example). Dimensional consistency implies that the product in each term of β_{λ_i} has the same dimension as λ_i, and that the product in each term of $\beta_{s_\alpha(x)}$ has the same dimension as $s_\alpha(x)$. The generalization is

$$\left(\mu\frac{\partial}{\partial\mu}+\sum_i\beta_{\lambda_i}\frac{\partial}{\partial\lambda_i}+\sum_\alpha\int\beta_{s_\alpha(x)}\frac{\partial}{\partial s_\alpha(x)}\,d^4x\right)W(\lambda,s)=0\,. \tag{1a.1.24}$$

In words, (1a.1.24) means that a change in μ can always be compensated by suitable changes in all the parameters **and** the fields.

Of course this assumes, as usual, that we have included all the parameters required to absorb the dependence on the physics at very short distances. Once all the relevant parameters have been included, then μ is arbitrary, and (1a.1.24) is almost a tautology.

In the simple example, (1a.1.1), the field, ϕ, is multiplicatively renormalized, because it is the only field around. In our renormalization group language, that statement is equivalent to (1a.1.23). Note, in fact, that (1a.1.12) is the most general thing we can write down consistent with dimensional analysis. However, in general, the situation is more complicated.

Suppose, for example, that your theory depends on several real scalar fields, ϕ_α, for $\alpha=1$ to K, with Lagrangian

$$\begin{aligned}\mathcal{L}(\phi)&=\frac{1}{2}\sum_\alpha\left(\partial^\mu\phi_\alpha\partial_\mu\phi_\alpha+s_\alpha\phi_\alpha\right)\\&-\sum_{\alpha_1\alpha_2\alpha_3\alpha_4}\frac{\lambda_{\alpha_1\alpha_2\alpha_3\alpha_4}}{4!}\,\phi_{\alpha_1}\phi_{\alpha_2}\phi_{\alpha_3}\phi_{\alpha_4}\,,\end{aligned} \tag{1a.1.25}$$

where $\lambda_{\alpha_1\alpha_2\alpha_3\alpha_4}$ is some symmetric tensor in the α's. If all $\frac{(K+3)!}{4!(K-1)!}$ of the independent α's are nonzero, then there are no symmetries that distinguish one ϕ from another. In this case, each of the β_λ-functions depends on all of the λ's, and the most general form for β_s is

$$\beta_{s_\alpha(x)}=\sum_{\alpha'}s_{\alpha'}(x)\,\gamma_{\alpha'\alpha}(\lambda)\,. \tag{1a.1.26}$$

This exhibits the phenomenon of operator mixing. Of course, all of the γ's are zero in one loop in the scalar field theory, but in higher order, a change in μ requires not just a rescaling of each of the ϕ_α's, but a rotation in the flavor space, to get the theory back into a canonical form. The fields are not multiplicatively renormalized.

The failure of multiplicative renormalization in (1a.1.26) is an example of the general principle that whatever can happen will happen. This is particularly true in quantum field theory, where if you fail to write down an appropriately general expression, the theory is likely to blow up in your face, in the sense that you will not be able to absorb the dependence on short distance physics. However, as long as you are careful to include everything (consistent, as we will emphasize time and time again, with all the symmetries of the system), the situation is really no more difficult to understand than in a simple theory with multiplicative renormalization. You just have to keep track of some indices.

As an example of operator mixing in action, we will show how to go back and forth from 1PI 2-point functions to connected 2-point functions. Consider the connected 2-point function

$$\begin{aligned}G_{\alpha_1\alpha_2}(x_1,x_2)&=-\frac{\delta^2}{\delta s_{\alpha_1}(x_1)\delta s_{\alpha_2}(x_2)}W(s)\\&=\langle 0\,|\,T\,\phi_{\alpha_1}(x_1)\phi_{\alpha_2}(x_2)\,|\,0\rangle\,.\end{aligned} \tag{1a.1.27}$$

Because $W(s)$ satisfies, (1a.1.24), we can differentiate twice and use (1a.1.26) to obtain with

$$\left(\mu\frac{\partial}{\partial\mu}+\beta_\lambda\frac{\partial}{\partial\lambda}\right)\equiv\mathcal{D}\,, \tag{1a.1.28}$$

$$\sum_{\alpha_1'\alpha_2'}\left[\mathcal{D}\,\delta_{\alpha_1\alpha_1'}\,\delta_{\alpha_2\alpha_2'}+\gamma_{\alpha_1\alpha_1'}+\gamma_{\alpha_2\alpha_2'}\right]\,G_{\alpha_1'\alpha_2'}(x_1,x_2)=0\,. \tag{1a.1.29}$$

Let $\tilde{G}(p)$ be the Fourier transform of $G(x,0)$. To avoid getting lost in indices, it will be useful to write (1a.1.29) for $\tilde{G}$ (Fourier transforming commutes with applying $\mathcal{D}$) in an obvious matrix notation,

$$\mathcal{D}\tilde{G}+\gamma\tilde{G}+\tilde{G}\gamma^T=0\,. \tag{1a.1.30}$$

The 1PI two point function is the inverse of $\tilde{G}$ — symbolically

$$\Gamma\tilde{G}=\tilde{G}\Gamma=I\,. \tag{1a.1.31}$$

Now apply $\mathcal{D}$ to (1a.1.31).

$$\begin{aligned}(\mathcal{D}\Gamma)\,\tilde{G}+\Gamma\,(\mathcal{D}\tilde{G})&=0\\ =(\mathcal{D}\Gamma)\,\tilde{G}-\Gamma\,\gamma\tilde{G}-\Gamma\,\tilde{G}\gamma^T&\ \cdot\\ =(\mathcal{D}\Gamma)\,\tilde{G}-\Gamma\,\gamma\tilde{G}-\gamma^T&\end{aligned} \tag{1a.1.32}$$

Multiplying by Γ on the right and using (1a.1.31) gives the renormalization group equation for the 1PI 2-point function:

$$\mathcal{D}\Gamma-\Gamma\gamma-\gamma^T\,\Gamma=0\,. \tag{1a.1.33}$$

Note not only the sign change, compared to (1a.1.30), but the change from γ to γ^T as well.

$\Gamma(\Phi)$ and 1PI Graphs

We can get directly from (1a.1.22) to a similarly general expression for $\Gamma(\Phi)$, the generating functional for 1PI graphs. $\Gamma(\Phi)$ is obtained by making a Legendre transformation on $W(s)$,

$$\Gamma(\Phi)=W(s)-\sum_\alpha\int s_\alpha(x)\,\Phi_\alpha(x)\,d^4x\,, \tag{1a.1.34}$$

where

$$\Phi_\alpha(x)=\frac{\delta W}{\delta s_\alpha(x)}\,, \tag{1a.1.35}$$

is the "classical field" corresponding to the quantum field, ϕ_α.

Now

$$\left(\mu\frac{\partial}{\partial\mu}+\beta_\lambda\frac{\partial}{\partial\lambda}\right)\Gamma(\Phi)=\left(\mu\frac{\partial}{\partial\mu}+\beta_\lambda\frac{\partial}{\partial\lambda}\right)W(s)\,, \tag{1a.1.36}$$

for fixed Φ. We don't have to worry about any implicit dependence of s on μ or λ, because the contribution from the two terms on the right hand side cancel anyway — the Legendre transform guarantees that $\Gamma(\Phi)$ doesn't depend on s. But the right hand side of (1a.1.36) is (from (1a.1.24))

$$-\sum_\alpha\int\beta_{s_\alpha(x)}\frac{\partial}{\partial s_\alpha(x)}\,W(\lambda,s)\,d^4x\,. \tag{1a.1.37}$$

But

$$s(x) = -\frac{\delta\Gamma}{\delta\Phi}, \tag{1a.1.38}$$

which follows from (1a.1.34). Putting this into (1a.1.23) and using (1a.1.35), we find

$$\begin{aligned} &\left(\mu\frac{\partial}{\partial\mu} + \beta_\lambda\frac{\partial}{\partial\lambda} - \sum_{\alpha,\alpha'} \Phi_\alpha\,\gamma_{\alpha'\alpha}\,\frac{\partial}{\partial\Phi_{\alpha'}}\right)\Gamma(\Phi) \\ &= \left(\mu\frac{\partial}{\partial\mu} + \beta_\lambda\frac{\partial}{\partial\lambda} - \sum_{\alpha,\alpha'} \Phi_\alpha\,\gamma^T_{\alpha\alpha'}\,\frac{\partial}{\partial\Phi_{\alpha'}}\right)\Gamma(\Phi) = 0\,. \end{aligned} \tag{1a.1.39}$$

From this, (1a.1.33) follows immediately by differentiation twice with respect to Φ_α.

An Example with Fermions

Consider the following Lagrangian, describing K pseudoscalar fields, and a massless Dirac fermion:

$$\begin{aligned} \mathcal{L}(\phi,\psi) &= i\overline{\psi}\,\partial\!\!\!/\,\psi + \overline{\eta}\psi + \overline{\psi}\eta + \sum_\alpha\left(\frac{1}{2}\partial^\mu\phi_\alpha\partial_\mu\phi_\alpha + s_{\phi_\alpha}\phi_\alpha\right) \\ &- \sum_{\alpha_1\alpha_2\alpha_3\alpha_4}\frac{\lambda_{\alpha_1\alpha_2\alpha_3\alpha_4}}{4!}\,\phi_{\alpha_1}\phi_{\alpha_2}\phi_{\alpha_3}\phi_{\alpha_4} - \sum_\alpha g_\alpha\,\overline{\psi}\,i\gamma_5\phi_\alpha\,\psi\,, \end{aligned} \tag{1a.1.40}$$

The quantum fermion fields, ψ, have dimension 3/2. Their sources, η, have dimension 5/2. Both are anticommuting.

The Lagrangian, (1a.1.40), is the most general Lagrangian we can write down for massless fields with these dimensions, consistent with a parity symmetry under which the ϕ_α fields are odd. Thus, (1a.1.24) will be satisfied, where the couplings now include the g_α's as well as the λ's, and the sources include the η and $\overline{\eta}$, as well as the s_α's. A feature of this theory that makes it a nice pedagogical example, is that all β-functions are nonzero at the one loop level. You will find all of these for yourselves in problem (1a-3).

1a.2 Composite Operators

In local field theory, it is useful to discuss new local objects that are built out of the local fields themselves. These things are called "composite operators". One reason for considering them, of course, is that the Lagrangian itself is built out of such things, but many other reasons will occur to us as we go along. The question is, what do we mean by multiplying the fields together at the same space time point. This is the same problem we faced in defining the theory in the first place. It is most sensible to deal with it in precisely the same way.

1. Couple a source, t, to each composite operator in the Lagrangian. We will find that if we include an operator of a given dimension, we will be forced to include sources for all the other operators with the same dimension and the same quantum numbers under all the symmetries of the theory. When masses are present, we may have to include others as well, but the theory will complain if we do not have the right sources.

2. Renormalize the theory as usual. To do this, you will have to include extra parameters for the new "interactions" that you can build with your new sources, one for each new dimension 4 object you can build. These will be required to absorb dependence on the unknown short distance physics.

3. Once this is done, the generating functional, W, depends on the new sources, and we can find the effect of an insertion of a composite operator in any Green function simply by differentiating with respect to the appropriate source.

As a fairly simple example, but one which exhibits many of the interesting features of composite operators, consider the fermion example, (1a.1.40), with $K = 1$ (dropping the unneeded subscripts), and include a source for the operator $\overline{\psi}\psi$. This is the only dimension 3 operator that is a scalar under parity, so we do not need any other sources. The Lagrangian, constructed according to the above rules, is

$$\begin{aligned}\mathcal{L}(\phi,\psi) = i\overline{\psi}\partial\!\!\!/\psi + \overline{\eta}\psi + \overline{\psi}\eta + \frac{1}{2}\partial^\mu\phi\partial_\mu\phi + s_\phi\phi \\ -\frac{\lambda}{4!}\phi^4 - g\overline{\psi}i\gamma_5\phi\psi + t\overline{\psi}\psi - \frac{\kappa_1}{2}\partial^\mu t\partial_\mu t - \frac{\kappa_2}{4!}t^4 - \frac{\kappa_3}{4}t^2\phi^2\,,\end{aligned} \tag{1a.2.1}$$

The point of the extra terms in (1a.2.1) is that they are require for the calculation of graphs involving more than one insertion of the new operator. For example, consider the graph

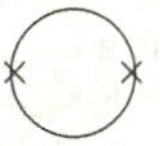

(1a.2.2)

where the $\times$ indicates the insertion of the operator $\overline{\psi}\psi$. When computed using DRMS, the graph is μ-dependent. This means that it depends on the details of how we constructed the $\overline{\psi}\psi$ operator at short distances, so we have no right to calculate it. Thus we need a new, *a priori* unknown parameter in the theory. This is κ_1.

The κ parameters have β-functions just like the other coupling constants. The μ-dependence of (1a.2.2) produces μ-dependence in κ_1, which is just its β-function in the renormalization group. If, somehow, we manage to discover the values of this and the other κ parameters at some μ (this requires somehow specifying the physics — we will see some examples later on), then we can use the renormalization group to find its value at any other μ.

The κ parameters are the translation into our language of so-called "subtractions" which must be made to define the Green functions of composite operators. We now see that they are something that we already understand. The sources for the composite fields differ from the sources for elementary fields in that their dimensions are smaller. This is what allows us to build the nontrivial κ terms. We will see many examples of composite operators as we explore the standard model. The thing to remember is that in this way of looking at things, they are not so different from other fields and sources.

Problems

1a-1. Find the one-loop β-functions for (1a.1.25). If you use (1a.1.19), no calculation should be required, beyond drawing the graphs and keeping track of the indices.

1a-2. Show that (1a.1.21) follows from (1a.1.22).

1a-3. Find all the β-functions, both for the couplings and for the sources, for the theory described by (1a.1.40), for $K = 2$.

1b — Gauge Symmetries

In this section, we will discuss theories with local or gauge symmetry. First, we will use gauge symmetry as a device to define the Noether currents in a quantum field theory. Then we will go on the discuss dynamical gauge symmetry.

Given a symmetry of a classical Lagrangian without sources, we can always extend it to a symmetry of the Lagrangian including sources, by requiring the sources to transform appropriately. The interesting question is, can we always define a corresponding **quantum** theory that exhibits the symmetry.

The answer is — sometimes! When it is possible to find a regularization and renormalization scheme that preserves the symmetry, then the symmetry can be extended into the quantum theory, but sometimes this is not possible. We will discuss important examples of the "anomalies" that can occur in quantum field theory to prevent the realization of a classical symmetry later. For now we will ignore this subtlety, and discuss symmetry in the language we have developed to discuss field theory, without asking whether any particular symmetry is consistent with our DRMS scheme.

1b.1 Noether's Theorem – Field Theory

Consider, then, a quantum field theory with N real scalar fields, ϕ, and real sources s and with K fermion fields, ψ, and sources, η, and Lagrangian[1]

$$\mathcal{L}_0(\phi, \partial^\mu\phi, \psi, \partial^\mu\psi) + s^T\,\phi + \overline{\eta}\,\psi + \overline{\psi}\,\eta\,, \tag{1b.1.1}$$

Suppose that $\mathcal{L}_0$ is invariant under a global internal symmetry group, G, as follows:

$$\begin{aligned} \phi &\rightarrow \mathcal{D}_\phi(g^{-1})\,\phi\,, \\ \psi_L &\rightarrow \mathcal{D}_L(g^{-1})\,\psi_L\,, \\ \psi_R &\rightarrow \mathcal{D}_R(g^{-1})\,\psi_R\,, \end{aligned} \tag{1b.1.2}$$

where g is and element of G and the $\mathcal{D}$'s are unitary representations of the symmetry group. Note that because the ϕ fields are real, the representation $\mathcal{D}_\phi$ must be a "real" representation. That is, the matrices, $\mathcal{D}_\phi$, must be real. Such a symmetry can then be extended to include the sources as well, as follows:

$$\begin{aligned} s &\rightarrow \mathcal{D}_\phi(g^{-1})\,s\,, \\ \eta_R &\rightarrow \mathcal{D}_L(g^{-1})\,\eta_R\,, \\ \eta_L &\rightarrow \mathcal{D}_R(g^{-1})\,\eta_L\,. \end{aligned} \tag{1b.1.3}$$

---

[1] We use $\mathcal{L}_0$ for the Lagrangian without any sources.

Note the switching of $L \leftrightarrow R$ for the fermion sources, because the fermions source terms are like mass terms that couple fields of opposite chirality. Now if the regularization and renormalization respect the symmetry, the vacuum amplitude, $Z(s, \eta, \overline{\eta})$ will also be invariant under (1b.1.3).

Generators — T^a

It is often convenient to use the infinitesimal version of (1b.1.2) and (1b.1.3), in terms of the generators, T^a,

$$\begin{aligned}
\delta\phi &= i\epsilon_a T^a_\phi\, \phi\,, \\
\delta\psi_L &= i\epsilon_a T^a_L\, \psi_L\,, \\
\delta\psi_R &= i\epsilon_a T^a_R\, \psi_R\,, \\
\delta s &= i\epsilon_a T^a_\phi\, s\,, \\
\delta\eta_R &= i\epsilon_a T^a_L\, \eta_R\,, \\
\delta\eta_L &= i\epsilon_a T^a_R\, \eta_L\,.
\end{aligned} \tag{1b.1.4}$$

The T^a_j for $j = \phi$, L and R, are generators of the representation of the Lie algebra of G, corresponding to the representations,

$$D_j(g^{-1}) = e^{i\epsilon_a T^a_j}\,. \tag{1b.1.5}$$

A sum over repeated a indices is assumed. Often we will drop the subscript, and let the reader figure out from the context which representation we are discussing.

Gauge Symmetry

In order to discuss the conserved Noether currents associated with a symmetry of the quantum field theory as composite operators, it is useful to convert the global symmetry, (1b.1.4), into a gauge symmetry, that is a symmetry in which the parameters can depend on space-time,

$$\epsilon_a \to \epsilon_a(x)\,. \tag{1b.1.6}$$

To do this, we must introduce a set of "gauge fields", which in this case will be classical fields, like the other sources. Call these fields, h^μ_a. Under the symmetry, (1b.1.4), they transform as follows:

$$\delta h^\mu_a = -f_{abc}\epsilon_b h^\mu_c + \partial^\mu \epsilon_a\,. \tag{1b.1.7}$$

Now we modify the Lagrangian, (1b.1.1), by replacing the derivatives by "covariant" derivatives,

$$D^\mu \equiv \partial^\mu - i\, h^\mu_a T^a\,, \tag{1b.1.8}$$

where the T^a are the generators of the representation of the object on which the derivative acts. Thus,

$$\begin{aligned}
D^\mu\phi &\equiv \left(\partial^\mu - i\, h^\mu_a T^a_\phi\right)\phi\,, \\
D^\mu\psi &\equiv \left(\partial^\mu - iP_+\, h^\mu_a T^a_L - iP_-\, h^\mu_a T^a_R\right)\psi\,.
\end{aligned} \tag{1b.1.9}$$

The resulting Lagrangian is now invariant under (1b.1.4) and (1b.1.7), for space-time dependent parameters, ϵ_a.

Furthermore, the classical gauge fields, h_a^μ, act as sources for composite fields in the quantum field theory. These are the classical Noether currents corresponding to the global symmetry, (1b.1.2). To see this, note that

$$\frac{\delta D^\mu \phi}{\delta h_a^\nu} = -i\,\delta_\nu^\mu\, T_\phi^a\, \phi \tag{1b.1.10}$$

Suppose that the Lagrangian depended only on the ϕs. Then using (1b.1.10), the chain rule for functional differentiation and the fact that ϕ doesn't depend on h_a^μ, we could write

$$\frac{\delta \mathcal{L}}{\delta\, h_a^\mu} = \frac{\delta \mathcal{L}}{\delta\, D^\nu \phi}\,\frac{\delta\, D^\nu \phi}{\delta\, h_a^\mu} = -i \frac{\delta \mathcal{L}}{\delta\, D^\mu \phi}\, T_\phi^a\, \phi \tag{1b.1.11}$$

which is the classical Noether current.

This works in general, because for any field ξ,

$$\frac{\delta\, D^\mu \xi}{\delta h_a^\nu} = -i\,\delta_\nu^\mu\, T_\xi^a\, \xi \tag{1b.1.12}$$

where T_ξ^a are the generators of the symmetry acting on the field ξ. Thus

$$\frac{\delta \mathcal{L}}{\delta\, h_a^\mu} = \sum_\xi \frac{\delta \mathcal{L}}{\delta\, D^\nu \xi}\,\frac{\delta\, D^\nu \xi}{\delta\, h_a^\mu} = -i \sum_\xi \frac{\delta \mathcal{L}}{\delta\, D^\mu \xi}\, T_\xi^a\, \xi \tag{1b.1.13}$$

which is the general form of the classical Noether current.

We will find that this definition of the Noether currents as the composite fields coupled to external gauge fields is very convenient because gauge invariance is such a strong constraint. The conservation of the Noether currents implies a set of relations for the Green functions known as Ward identities. All of these follow from the gauge invariance of the vacuum functional, $Z(\phi, \eta, \overline{\eta}, h_a^\mu)$.

Note that because the field h_a^μ has dimension 1, like a derivative, there is one kind of term of dimension 4 that we can write, consistent with all the symmetries, that depends only on h_a^μ. It is

$$\frac{\kappa_{ab}}{4}\, h_a^{\mu\nu}\, h_{b\mu\nu}\,, \tag{1b.1.14}$$

where $h_a^{\mu\nu}$ is the "gauge field strength",

$$h_a^{\mu\nu} = \partial^\mu h_a^\nu - \partial^\nu h_a^\mu + f_{abc}\, h_b^\mu h_c^\nu\,. \tag{1b.1.15}$$

Thus we expect such terms to be required by renormalization. With our convenient normalization, (1.1.12), the parameter κ_{ab} is

$$\kappa_{ab} = \delta_{ab}\, \kappa_a \tag{1b.1.16}$$

where κ_a is independent of a within each simple subalgebra.[2]

Some Useful Notation

It is often useful to get rid of the indices on the gauge field and replace the set by a matrix field

$$h^\mu \equiv T^a\, h_a^\mu\,. \tag{1b.1.17}$$

The covariant derivative is

$$D^\mu \equiv \partial^\mu - i\, h^\mu\,. \tag{1b.1.18}$$

[2]For $U(1)$ factors of the symmetry algebra, there is no normalization picked out by the algebra, and no constraint of the form (1b.1.16).

The field strength is

$$h^{\mu\nu} \equiv T^a\, h_a^{\mu\nu} = \partial^\mu h^\nu - \partial^\nu h^\mu - i\,[h^\mu, h^\nu]\,. \tag{1b.1.19}$$

The field strength can be obtained by commuting the covariant derivatives,

$$[D^\mu, D^\nu] = -i\, h^{\mu\nu}\,. \tag{1b.1.20}$$

In this notation, the gauge transformation, (1b.1.7), becomes

$$\delta h^\mu = i[\epsilon, h^\mu] + \partial^\mu \epsilon\,, \tag{1b.1.21}$$

where $\epsilon = \epsilon_a T_a$. In this form, the gauge transformation can be integrated to give the finite form

$$h^\mu \to \Omega h^\mu \Omega^{-1} + i\Omega \partial^\mu \Omega^{-1}\,, \tag{1b.1.22}$$

where

$$\Omega = e^{i\epsilon}\,. \tag{1b.1.23}$$

1b.2 Gauge Theories

In this section, we consider what happens when we let the classical gauge field, $h^{\mu\nu}$, becomes a quantum field,

$$\begin{aligned} h^\mu &\to G^\mu\,, \\ h^{\mu\nu} &\to G^{\mu\nu} \\ &= \partial^\mu G^\nu - \partial^\nu G^\mu - i\,[G^\mu, G^\nu]\,. \end{aligned} \tag{1b.2.1}$$

The κ term of (1b.1.14) becomes the kinetic energy term for the gauge fields,

$$-\sum_a \frac{1}{4g_a^2}\, G_{a\mu\nu} G_a^{\mu\nu}\,. \tag{1b.2.2}$$

Again, the normalization, that we have here called $1/4g_a^2$, is equal for a in each simple component of the gauge group. The constant, g_a, is the gauge coupling constant. We can rescale the fields to make the coefficient of the kinetic energy term the canonical 1/4 if we choose. Then the gauge transformation law has g_a dependence. Often, we will keep the form of the gauge transformations fixed and let the gauge coupling constant appear as the normalization of the kinetic energy term. This is particularly convenient for the study of the symmetry structure of the theory.

Gauge Fixing

The field theory treatment of gauge theories is more complicated than that of theories without dynamical gauge symmetry because the gauge symmetry must be broken in order to define the theory, at least perturbatively. The kinetic energy term, (1b.2.2), does not involve the longitudinal components of the gauge fields, and is not invertible in momentum space. Thus the propagator is not well defined.

In the functional integral formulation of field theory, one tries to write the Green functions as functional integrals:

$$\langle 0 \,|\, T\, g(G,\phi) \,|\, 0 \rangle = \frac{\langle \tilde{g} \rangle}{\langle \tilde{1} \rangle}\,, \tag{1b.2.3}$$

where

$$\langle \tilde{g} \rangle = \int g(G,\phi)\, e^{iS(G,\phi)}\, [dG][d\phi]\,. \tag{1b.2.4}$$

In the functional integral description, the difficulty with gauge invariance shows up in the independence of the action on gauge transformation,

$$G^\mu \to \Omega G^\mu \Omega^{-1} + i\Omega \partial^\mu \Omega^{-1}\,. \tag{1b.2.5}$$

This defines an "orbit" in the space of gauge field configurations on which the action is constant. The part of the gauge field functional integral over the gauge transformations is completely unregulated, and gives rise to infinities that are the translation into the functional integral language of the problem with the gauge field propagator. Thus $\langle \tilde{g} \rangle$ has spurious infinities, and is not well defined.

Having formulated the problem in this language, Fadeev and Popov suggested a general procedure for solving it. They introduced 1 into (1b.2.3), in the following form:

$$1 = \Delta(G,\phi) \int f(G_\Omega, \phi_\Omega)\, [d\Omega]\,, \tag{1b.2.6}$$

where f is a functional that is **not** gauge invariant and $[d\Omega]$ is the invariant measure over the gauge transformations, satisfying

$$[d\Omega] = [d(\Omega'\Omega)] = [d(\Omega\Omega')]\,, \tag{1b.2.7}$$

for fixed Ω'. The functional, Δ, called the Fadeev-Popov determinant, is gauge invariant. This can be seen as follows (using (1b.2.7)):

$$\Delta(G_\Omega, \phi_\Omega) = \frac{1}{\int f(G_{\Omega\Omega'}, \phi_{\Omega\Omega'})}\, [d\Omega'] \tag{1b.2.8}$$

$$\frac{1}{\int f(G_{\Omega\Omega'}, \phi_{\Omega\Omega'})\, [d(\Omega\Omega')]} = \frac{1}{\int f(G_\Omega, \phi_\Omega)\, [d(\Omega)]} = \Delta(G,\phi)\,. \tag{1b.2.9}$$

Inserting (1b.2.6), we can write

$$\langle \tilde{g} \rangle = \int e^{iS(G,\phi)}\, g(G,\phi)\, \Delta(G,\phi)\, f(G_\Omega, \phi_\Omega)\, [dG][d\phi][d\Omega]\,. \tag{1b.2.10}$$

If $g(G,\phi)$ is gauge invariant, then (1b.2.10) can be written with the integral over the gauge transformations factored out, as

$$\langle \tilde{g} \rangle = \left\{ \int e^{iS(G,\phi)}\, g(G,\phi)\, \Delta(G,\phi)\, f(G_\Omega, \phi_\Omega)\, [dG][d\phi] \right\} \int [d\Omega]\,. \tag{1b.2.11}$$

The Fadeev-Popov suggestion is then to drop the factor of $\int [d\Omega]$, and define

$$\langle g \rangle = \int e^{iS(G,\phi)}\, g(G,\phi)\, \Delta(G,\phi)\, f(G_\Omega, \phi_\Omega)\, [dG][d\phi]\,. \tag{1b.2.12}$$

This process is called "choosing a gauge". The gauge is defined by the choice of the fixing function, f. The important point to notice is that $\langle g \rangle$ should be completely independent of the gauge fixing function, $f(G,\phi)$, so long as the Green function, $g(G,\phi)$, is gauge invariant. Because it is the gauge invariant physical matrix elements that we actually want, this is a perfectly reasonable definition.

Of course, this means that in general, Green functions are not invariant for a given gauge fixing function, f. That means that the consequences of gauge symmetry are not as obvious as those of a global symmetry. However, it is possible to organize the calculations in gauge theory in a way that makes gauge invariance manifest, by using what is called "background field gauge". This is discussed in Appendix B.

1b.3 Global Symmetries of Gauge Theories

Let us look again at the Lagrangian for a general gauge theory,

$$\mathcal{L}_0(\phi, D^\mu\phi, \psi, D^\mu\psi) - \sum_a \frac{1}{4g_a^2} G_{a\mu\nu} G_a^{\mu\nu} \,, \tag{1b.3.1}$$

and study its global symmetries. Of course, the gauge symmetry itself defines a global symmetry, but the theory may have additional symmetry. We will reserve the term "flavor symmetry" to refer only to those internal symmetries of the theory that are not gauged. The first important comment is that any such symmetry must map the gauge fields into themselves. Because the gauge fields are associated with the generators of the gauge symmetry, this implies that the gauge symmetry is a normal subgroup of the full symmetry group. For continuous flavor symmetries, the constraint is even stronger. The generators of continuous flavor symmetries must commute with all the gauge generators.[3] This, in turn, implies that the flavor generators are constant within each irreducible representation of the gauge group, and also that if two fields are rotated into one another by the flavor symmetry, they must belong to the same irreducible representation of the gauge group. If we choose a canonical form for the gauge generators in each irreducible representation of the gauge symmetry,[4] then we can regard the ψ and ϕ fields as having two sets of indices: gauge indices, on which the gauge symmetries act; and flavor indices, on which the flavor symmetries act.

First consider the fermion kinetic energy term. In the basis in which all the fermions are left handed, this has the general form

$$i \sum_r \overline{\psi}_{rL} \not{D} \psi_{rL} \,, \tag{1b.3.2}$$

where the sums run over the different irreducible representations of the gauge group, and where, for each r, the fermions field is a vector in a flavor space with dimension d_r. Each of these flavor vectors can be rotated without changing (1b.3.2). Thus, at least classically, (1b.3.2) has a separate $SU(d_r) \times U(1)$ flavor symmetry for each irreducible representation that appears in the sum (with non-zero d_r).[5]

If we charge conjugate some of the left handed fermion fields to write them as right handed fields, we must remember that the charged conjugate fields may transform differently under the gauge transformations. If

$$\delta\psi_L = i\epsilon_a T^a \psi_L \,, \tag{1b.3.3}$$

then

$$\delta\psi_{cR} = i\epsilon_a(-T^{a*})\psi_{cR} \,, \tag{1b.3.4}$$

In a "real" representation, the generators can be taken to be imaginary, antisymmetric matrices, so that $-T^{a*} = T^a$, and the right handed fields transform the same way. However, if the representation is "complex", then $-T^{a*}$ is a different representation. Finally, the representation may be "pseudo-real", in which case the representation generated by $-T^{a*}$ is equivalent to that generated by T^a,

$$-T^{a*} = S\, T^a\, S^{-1} \,, \tag{1b.3.5}$$

but the matrix S is antisymmetric, so the generators can not be made entirely real. In this case, one can include the matrix S into the definition of charge conjugation to make the right handed fields

[3] Discrete symmetries, like parity, my induce nontrivial automorphisms on the gauge generators.

[4] See problem (3-2) for an example of what happens when you do not do this.

[5] As we will see later, some of the $U(1)$s are broken by quantum mechanical effect, "anomalies".

transform like the left handed fields. The most familiar example of a pseudo-real representation is the Pauli matrices, generating the spin 1/2 representation of $SU(2)$.

The situation with scalar fields is somewhat more subtle, because the global symmetry associated with a given representation of the gauge group depends on whether it is real, complex, or pseudo-real.

For d sets of scalars in a real irreducible representation, r_R, of the gauge group, the flavor symmetry is $SO(d)$. Here the kinetic energy term looks like

$$D^\mu \Phi^T D_\mu \Phi \,, \tag{1b.3.6}$$

where Φ is a vector in flavor space (as well as the space of the gauge group representation).

If, instead, the scalar field representation, r_C, is complex, then because the elementary scalar fields are real, both the r_C representation and its complex conjugate, $\overline{r}_C$, appear in the theory. Suppose that there are d copies of the representation, r_C, which we can organize into a vector in flavor space, as usual. We can write the kinetic energy term as

$$D^\mu \phi^\dagger D_\mu \phi \,. \tag{1b.3.7}$$

However, (1b.3.7) describes the $\overline{r}_C$ representation as well, because we could just as well have used the field ϕ^*, which transforms under the $\overline{r}_C$ representation. If the representation, r_C (and $\overline{r}_C$) is n dimensional, then there are actually $2nd$ real fields described by (1b.3.7), because each of the components of ϕ is complex. The flavor symmetry of (1b.3.7) is an $SU(d)$ acting on the flavor indices of ϕ.

The most bizarre situation occurs for pseudo-real representations. It turns out that a pseudo-real representation has an $SU(2)$ structure built into it. The generators of a $2n$ dimensional irreducible pseudo-real representation can be taken to have the form[6]

$$T_a = A_a + \vec{\tau} \cdot \vec{S}_a \,, \tag{1b.3.8}$$

where $\vec{\tau}$ are the Pauli matrices, and the $n \times n$ matrix A is antisymmetric while the $n \times n$ $\vec{S}$ matrices are symmetric. Then

$$\tau_2 T_a^* \tau_2 = -T_a \,. \tag{1b.3.9}$$

The flavor structure of the theory involves the $\vec{\tau}$ matrices in a non-trivial way. If there are d identical representations, we can describe them by a $2n \times 2d$ matrix field of the following form:

$$\begin{pmatrix} \Sigma_{11} & \cdots & \Sigma_{1d} \\ \vdots & \ddots & \vdots \\ \Sigma_{n1} & \cdots & \Sigma_{nd} \end{pmatrix} , \tag{1b.3.10}$$

where Σ_{ij} is a 2×2 matrix in the Pauli space, of the special form

$$\Sigma_{ij} = \sigma_{ij} + i\vec{\tau} \cdot \vec{\pi}_{ij} \,, \tag{1b.3.11}$$

for real σ_{ij} and $\vec{\pi}_{ij}$. The gauge generators act on Σ on the left. Now the kinetic energy term has the form

$$\text{tr}\left(D^\mu \Sigma^\dagger D_\mu \Sigma\right) . \tag{1b.3.12}$$

This is invariant under an $Sp(2d)$ flavor symmetry, generated by acting on Σ on the right with matrices of the form

$$T_x = A_x + \vec{\tau} \cdot \vec{S}_x \,, \tag{1b.3.13}$$

---

[6] Note that all pseudo-real representations have even dimensionality.

where the A and $\vec{S}$ are $d \times d$ matrices, antisymmetric and symmetric, respectively. Like the gauge transformation, (1b.3.13) preserves the special form of Σ.

Thus the largest possible flavor symmetry of a set of d pseudo-real representations of scalars is $Sp(2d)$. This apparently arcane fact will be important when we discuss the standard model with a fundamental Higgs boson.

Problems

1b-1. In the theory described by (1b.1.1), find the μ dependence of κ_{ab} in the one loop approximation. This does not involve any of the unspecified interactions, because it arises from one loops diagrams like (1a.2.2). It will depend on the representation matrices, T_f^a.

1b-2. Consider the Lagrangian

$$i\overline{\psi}_1 \not{D}_1 \psi_1 + i\overline{\psi}_2 \not{D}_2 \psi_2$$

where

$$D_j^\mu = \partial^\mu - iG_a T_j^a$$

with

$$T_1^a = \frac{1}{2}\tau_a$$

where τ_a are the Pauli matrices, but

$$T_2^1 = \frac{1}{2}\tau_2\,, \quad T_2^2 = \frac{1}{2}\tau_3\,, \quad T_2^3 = \frac{1}{2}\tau_1\,.$$

This theory has an SU(2) gauge symmetry, and also an SU(2) global flavor symmetry, because ψ_1 and ψ_2 transform under **equivalent** representations of the gauge symmetry which we have chosen, perversely, to write in different forms. Find the generators of the SU(2) flavor symmetry.

2 — Weinberg's Model of Leptons

2.1 The electron

Quantum electrodynamics (QED) is the quantum field theory of electrons and photons. It gives a spectacularly accurate description of the electron's properties in terms of only two parameters, the electron mass, m_e, and the fine structure constant, α. The success of QED derives from two special characteristics of the electron:

1. that it is the lightest charged particle, by a large factor;

2. it does not carry color and therefore does not participate directly in the strong interactions.

This second property is a defining characteristic of the class of spin 1/2 particles called "leptons" comprising the electron e^- and its neutrino ν_e; the muon μ^- and its neutrino ν_μ and the tau t^- and its neutrino ν_τ — and by their antiparticles. So far as we know today, the μ^- and the τ^- seem to be heavier copies of the electron, distinguished only by their larger masses. As far as we can tell, the weak *and* electromagnetic interactions act on the μ and τ exactly as they act on the electron. Because they are heavier, they decay, by weak interactions. The electron, the lightest charged particle, must be absolutely stable unless electromagnetic gauge invariance and global charge conservation are violated.

The $SU(2) \times U(1)$ theory of the weak and electromagnetic interactions was first written down as a model of leptons, simply because at the time the strong interactions the weak interactions of the hadrons were not completely understood.

2.2 $SU(2) \times U(1)$

In its simplest and original version, the $SU(2) \times U(1)$ model describes the weak interactions of the leptons. Nine fields needed to describe the e, μ, and τ and their neutrinos.

$$\nu_{eL}\,, \quad e_L^-\,, \quad e_R^- \tag{2.2.1}$$

$$\nu_{\mu L}\,, \quad \mu_L^-\,, \quad \mu_R^- \tag{2.2.2}$$

$$\nu_{\tau L}\,, \quad \tau_L^-\,, \quad \tau_R^- \tag{2.2.3}$$

There is no compelling evidence for right-handed ν's or left-handed $\overline{\nu}$'s, so we do not need a ν_R field.

The neutrinos are known to be very light. They may be massless, but there are some (still confusing) indications that they may have small masses. If the neutrinos are massless, the weak interactions conserve electron number, μ number and τ number separately. Formally, this means that the theory has global symmetries:

$$\begin{aligned} \nu_{eL} &\to e^{i\theta_e}\nu_{eL}\,, \quad e_L^- \to e^{i\theta_e}e_L^-\,, \quad e_R^- \to e^{i\theta_e}e_R^-\,; \\ \nu_{\mu L} &\to e^{i\theta_\mu}\nu_{\mu L}\,, \quad \mu_L^- \to e^{i\theta_\mu}\mu_L^-\,, \quad \mu_R^- \to e^{i\theta_\mu}\mu_R^-\,; \\ \nu_{\tau L} &\to e^{i\theta_\tau}\nu_{\tau L}\,, \quad \tau_L^- \to e^{i\theta_\tau}\tau_L^-\,, \quad \tau_R^- \to e^{i\theta_\tau}\tau_R^-\,. \end{aligned} \tag{2.2.4}$$

We will come back in later chapters to the lepton number violating effects that can occur if the neutrino masses are non-zero. For now, we will ignore neutrino masses. Then the symmetries, (2.2.4), simplify the construction of the model because the different families do not mix with one another.

Further, the interactions of the μ and τ are exact copies of those of the electron, so we can discuss only the fields in the electron family (2.2.1).

The gauge group is $SU(2) \times U(1)$, which means that there are four vector fields, three of which are associated with the $SU(2)$ group that we will call W_a^μ, where $a = 1$, 2 or 3, and one X_μ associated with the $U(1)$.

The structure of the gauge theory is determined by the form of the covariant derivative:

$$D^\mu = \partial^\mu + igW_a^\mu T_a + ig' X^\mu S \tag{2.2.5}$$

where T_a and S are matrices acting on the fields, called the generators of $SU(2)$ and $U(1)$, respectively. Notice that the coupling constants fall into two groups: there is one g for all of the $SU(2)$ couplings but a different one for the $U(1)$ couplings. The $SU(2)$ couplings must all be the same (if the T_a's are normalized in the same way $\kappa\ \mathrm{tr}(T_a T_b) = \delta_{ab}$) because they mix with one another under global $SU(2)$ rotations. But the $U(1)$ coupling g' can be different because the generator S never appears as a commutator of $SU(2)$ generators. Even if we did start with equal g and g', we would be unable to maintain the equality in the quantum theory in any natural way. The two couplings are renormalized differently and so require different infinite redefinitions in each order of perturbation theory. Since we need different counterterms, it would be rather silly to relate the couplings.

To specify the gauge structure completely, we must define the action of T_a and S on the fermion fields. Define the doublet

$$\psi_L \equiv \begin{pmatrix} \nu_{eL} \\ e_L^- \end{pmatrix} \tag{2.2.6}$$

Then the T_a's are defined as

$$T_a \psi_L = \frac{\tau_a}{2}\psi_L\,, \quad T_a e_R^- = 0 \tag{2.2.7}$$

where τ_a are the Pauli matrices. Both of these sets of T's satisfy the $SU(2)$ commutation relation

$$[T_a\,,\ T_b] = i\epsilon_{abc}T_c \tag{2.2.8}$$

although for the e_R^- field, which is called an $SU(2)$ singlet, (2.2.8) is satisfied in a rather trivial way.

In order to incorporate QED into the theory we are building, we must make certain that some linear combination of the generators is the electric-charge matrix Q. The matrix T_3 is clearly related to the charge because the difference between the T_3 values of each multiplet (the doublet ψ_L and, trivially, the singlet e_R^-) us the same as the charge difference. Thus, we define

$$Q = T_3 + S \tag{2.2.9}$$

which defines S. We have done this very carefully so that S will be proportional to the unit matrix on each multiplet.

$$Q\nu_{eL} = 0\,; \quad Qe^-_L = -e^-_L\,; \quad S\psi_L = -\frac{1}{2}\psi_L\,; \quad Qe^-_R = Se^-_R = -e^-_R\,. \tag{2.2.10}$$

We defined S this way so it satisfies the $SU(2) \times U(1)$ commutation relations

$$[T_a\,,\,S] = 0\,. \tag{2.2.11}$$

(2.2.5)-(2.2.11) completely define the gauge couplings to the leptons. To see just what these gauge couplings do, we will look first at the interactions that change particle identity, the couplings of W_1^μ and W_2^μ. If we write out the coupling of W_1^μ and W_2^μ to the fermions just by inserting the standard forms of the Pauli matrices into (2.2.5)-(2.2.7), we find the interaction terms

$$-\frac{g}{2}\left\{\overline{\nu_{eL}}(\not{W}_1 - i\not{W}_2)e^-_L + \overline{e^-_L}(\not{W}_1 + i\not{W}_2)\nu_{eL}\right\} \tag{2.2.12}$$

plus an analogous term for muons. The "charged" fields defined by

$$W^\mu_\pm = \frac{W_1 \mp iW_2}{\sqrt{2}} \tag{2.2.13}$$

create and annihilate charged intermediate vector bosons. $W^\mu_\pm$ annihilates (creates) $W^\pm$ ($W^\mp$) particles. (2.2.10) can give rise to μ decay as shown in Figure 2-1.

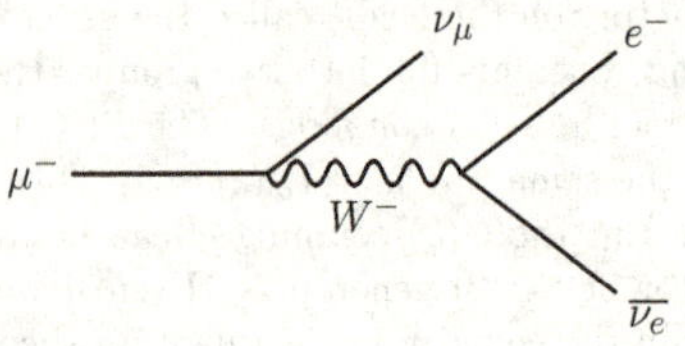

Figure 2-1:

There are two things wrong with Figure 2-1 as a picture of the weak interactions. The μ^- and e^- are massless, and the $W^\pm$ are massless. The $SU(2) \times U(1)$ gauge symmetry does not allow a lepton mass term $\overline{e^-_L}e^-_R$ or a $W^\pm$ mass term. The leptons must obviously get mass somehow for the theory to be sensible. The $W^\pm$ must also be very heavy in order for the theory to agree with data. A massless $W^\pm$ would give rise to a long-range weak force. In fact, the force has a very short range.

Despite these shortcomings, we will press on and consider the neutral sector. If the theory is to incorporate QED, one linear combination of the W_3^μ and X^μ fields must be the photon field A^μ. Thus, we write

$$A^\mu = \sin\theta\, W_3^\mu + \cos\theta\, X^\mu \tag{2.2.14}$$

Then the orthogonal linear combination is another field called Z^μ:

$$Z^\mu = \cos\theta\, W_3^\mu - \sin\theta\, X^\mu \tag{2.2.15}$$

The two independent fields are orthogonal linear combinations of W_3 and X because I have taken the kinetic-energy terms for the gauge fields to be

$$-\frac{1}{4}W_a^{\mu\nu}W_{a\mu\nu} - \frac{1}{4}X^{\mu\nu}X_{\mu\nu} \tag{2.2.16}$$

where

$$
\begin{aligned}
W^{\mu\nu} &= \partial^\mu W_a^\nu - \partial^\nu W_a^\mu - g\epsilon_{abc} W_b^\mu W_c^\nu \\
X^{\mu\nu} &= \partial^\mu X^\nu - \partial^\nu X^\mu
\end{aligned}
\quad . \tag{2.2.17}
$$

Thus, only the orthogonal combinations of W_3 and X have independent kinetic energy terms.

At this point, $\sin\theta$ is an arbitrary parameter. But if we insert (2.2.14-2.2.15) into the covariant derivative, we must obtain the photon coupling of QED. This determines the couplings of g and g' in terms of $\sin\theta$ and e.

Schematically, the couplings of the neutral gauge particles are

$$
g\not{W}_3 T_3 + g'\not{X} S \tag{2.2.18}
$$

Inserting (2.2.14)-(2.2.15) we get couplings

$$
\not{A}(g\sin\theta\, T_3 + g'\cos\theta\, S) + \not{Z}(g\cos\theta\, T_3 - g'\sin\theta\, S)\,. \tag{2.2.19}
$$

Since $\not{A}$ must couple to $eQ = e(T_3 + S)$ [from (2.2.9)], we must have

$$
g = \frac{e}{\sin\theta}\,, \quad g' = \frac{e}{\cos\theta} \tag{2.2.20}
$$

Then the Z couplings are

$$
\not{Z}(e\cot\theta\, T_3 - e\tan\theta\, S) = \not{Z}\frac{e}{\sin\theta\cos\theta}(T_3 - \sin^2\theta\, Q) \tag{2.2.21}
$$

The interesting thing about (2.2.21) is that the Z, unlike the photon, has non-zero couplings to neutrinos. Z exchange produces so-called neutral-current weak interactions such as $\nu_\mu e^-$ elastic scattering, shown in Figure 2-2.

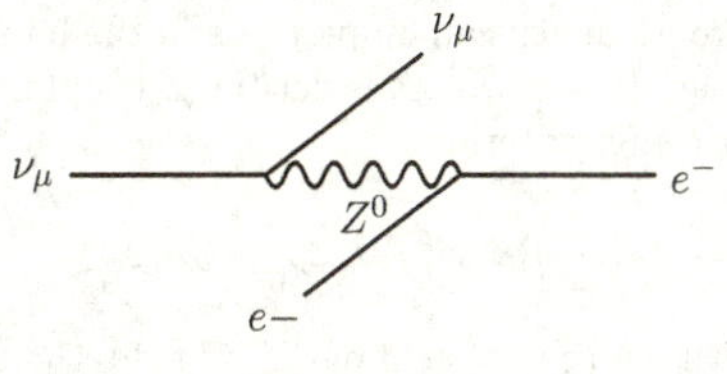

Figure 2-2:

The $SU(2)\times U(1)$ gauge structure was first written down by Glashow in 1960. At the time, the nature of the weak interactions was by no means obvious. That he got the right form at all was a great achievement that had to wait over 12 years for experimental confirmation. Of course, he did not know how to give mass to the $W^\pm$ and Z without breaking the gauge symmetry explicitly. The Z, like the $W^\pm$, must be very heavy. That much was known immediately because a massless Z would give rise to a peculiar parity-violating long-range force (see Figure 2-2.).

2.3 Renormalizability? An Interlude

So why don't we just add mass terms for the $W^\pm$ and Z by hand, even though it breaks the gauge symmetry? The problem is not lack of symmetry. Nice as the idea of local symmetry is, it is not what we care about. We are physicists, not philosophers. The real problem is that the theory with mass terms is not renormalizable, so we don't really know how to make sense of it. Before discussing massive vector bosons, let's go back an historical step and talk about the four-Fermi theory.

Before $SU(2) \times U(1)$, there was a phenomenological theory of the charged-current weak interactions based on a four-fermion interaction. For the μ decay interaction, the form was as follows:

$$\frac{G_F}{\sqrt{2}} J^\alpha_\mu J^*_{e\alpha} \tag{2.3.1}$$

where

$$J^\alpha_\mu = \overline{\nu_\mu} \gamma^\alpha (1+\gamma_5) \mu^- \tag{2.3.2}$$

and

$$J^\alpha_e = \overline{\nu_e} \gamma^\alpha (1+\gamma_5) e^- \tag{2.3.3}$$

(2.3.1) is called a current-current interaction for obvious reasons.

The constant G_F (F is for Fermi) is determined from the μ decay rate. It has units of $1/\text{mass}^2$

$$m_p^2 G_F \simeq 10^{-5} \tag{2.3.4}$$

where m_p is the proton mass.

(2.3.1) gives a perfectly adequate description of μ decay in tree approximation. The difficulty is that it is a dimension-6 operator, where the dimension of the fermion field is determined by the requirement that the kinetic-energy term $i\overline{\psi}\not{\partial}\psi$ has dimension 4. The theory is not renormalizable because quantum corrections produce an infinite sequence of interactions with higher and higher dimensions, all with infinite coefficients.

If we decide that we need a renormalizable theory (for example, we might regard the success of the e and μ $g-2$ predictions in QED as evidence for renormalizability in general), the four-Fermi theory is no good. We can do better by adding e^-, μ^-, and $W^\pm$ and Z terms to the gauge theory of Section 2.2.1. Then there are no dimension-6 operators in the interaction Hamiltonian. But the theory is still not renormalizable. To see this, consider the Z couplings (the $W^\pm$ would do as well).

The point is that the kinetic energy term is

$$-\frac{1}{4}(\partial^\mu Z^\nu)(\partial_\mu Z_\nu - \partial_\nu Z_\mu) \tag{2.3.5}$$

which doesn't involve the longitudinal component of the Z^μ field, the component proportional to the gradient of a scalar field. Let's call this longitudinal component Z_L and the transverse component Z_T.

$$Z^\mu = Z^\mu_T + Z^\mu_L \partial_\mu Z^\mu_T = 0, \quad \partial^\mu Z^\nu_L - \partial^\nu Z^\mu_L = 0\,. \tag{2.3.6}$$

The theory in which we add a Z mass term by hand, however, does involve Z_L in the mass term. Since Z_L appears only in the mass term, not in the kinetic-energy term, it acts like an auxiliary field. The Z_L propagator does not fall off with momentum. $M_L Z_L$ then acts like a field with dimension 2 for purposes of power counting. Then the Z_L couplings to other stuff, such as $\overline{\psi}\not{Z}_L\psi$ terms, are dimension 5 and not renormalizable.

But all is not lost. If we somehow preserve the gauge-invariance structure and give mass to the $W^\pm$ and Z, we may be able to preserve renormalizability. This is what Weinberg and Salam did by making use of spontaneous symmetry breaking. We will return to the general discussion of why (and whether) the theory should be renormalizable when we discuss effective field theories in Chapter 8.

2.4 Spontaneous Symmetry Breakdown

As a simple example of spontaneous symmetry breaking, consider the theory with a single Hermetian scalar field and the Lagrangian

$$\mathcal{L}(\phi) = \frac{1}{2}\partial^\mu\phi\partial_\mu\phi - \frac{m^2}{2}\phi^2 - \frac{\lambda}{4}\phi^4 \tag{2.4.1}$$

With only a single field, there can be no continuous internal symmetry, but this $\mathcal{L}(\phi)$ is invariant under the reflection

$$\phi \to -\phi \tag{2.4.2}$$

This has allowed us to omit a ϕ^3 (and ϕ) interaction term so that the theory has only two parameters, m and λ. If λ is small, we can hope to treat the $\lambda\phi^4$ term as a perturbation.

But suppose the sign of the mass term is changed so that the $\mathcal{L}(\phi)$ becomes (discarding a constant)

$$\begin{gathered}\mathcal{L}(\phi) = \tfrac{1}{2}\partial^\mu\phi\partial_\mu\phi - V(\phi) \\ V(\phi) = \tfrac{\lambda}{4}\left(\phi^2 - \tfrac{m^2}{\lambda}\right)^2\end{gathered} \tag{2.4.3}$$

In this case, we cannot perturb around the $\phi = 0$ vacuum because the free theory contains a tachyon, a particle with imaginary mass. Nor would we want to do so because the potential looks like what we see in Figure 2-3. It is clear that the $\phi = 0$ vacuum is unstable and that the theory much prefers to spend its time near one of two degenerate minima, $\phi = \pm\sqrt{m^2/\lambda}$. It doesn't care which.

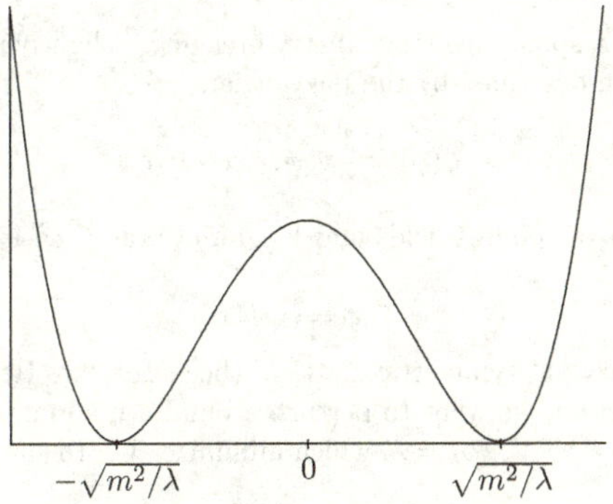

Figure 2-3:

If instead, we perturb around one of the minima, we get a sensible theory. The two are physically equivalent, so we pick $\phi = \sqrt{m^2/\lambda}$. For the ϕ field, $\sqrt{M^2/\lambda}$ is a "vacuum expectation value" (VEV). It is easier to see what is going on if we rewrite the theory in terms of a field with zero VEV.

$$\phi' = \phi - \sqrt{m^2/\lambda} \tag{2.4.4}$$

The result is

$$\begin{aligned} \mathcal{L}(\phi') &= \tfrac{1}{2}\partial^\mu\phi'\partial_\mu\phi' - V(\phi') \\ V(\phi') &= \tfrac{\lambda}{4}\left(\phi'^{\,2} + 2\sqrt{m^2/\lambda}\,\phi'\right)^2 \\ &= \tfrac{\lambda}{4}\phi'^{\,4} + \sqrt{m^2/\lambda}\,\phi'^{\,3} + m^2\phi'^{\,2} \end{aligned} \tag{2.4.5}$$

Several points should stand out. The symmetry is hidden. It has been spontaneously broken by the choice of vacuum. However, the theory is still described by only two parameters, for this relation between the $\phi'^{\,4}$, $\phi'^{\,3}$ and $\phi'^{\,2}$ terms will be preserved by quantum renormalization effects. This enhanced renormalizability, the fact that we need only two independent counterterms, is the legacy of the spontaneously broken symmetry. The ϕ' field describes a massive scalar field with mass $\sqrt{2}m$, with the original "imaginary" mass setting the scale, but not trivially related to the eventual physical particle masses.

The existence of the other possible vacuum, $\phi = -\sqrt{m^2/\lambda}$ or $\phi' = -2\sqrt{m^2/\lambda}$, does not show up in perturbation theory. It is infinitely far away because the field must be changed everywhere in space-time to get there. Note that spontaneous symmetry breaking occurs only in infinite space-time. In a finite space, the ground state would be a linear combination of the $\phi = \sqrt{m^2/\lambda}$ vacua, invariant under the discrete symmetry. In the infinite space, such states are forbidden by superselection rules. The existence of the other vacuum does give rise to a variety of interesting, and sometimes even more important, nonperturbative effects in the infinite volume theory, but we will ignore them for now and press on.

2.5 The Goldstone Theorem

For continuous symmetries, spontaneous symmetry breaking is slightly more subtle. Let us consider the fairly general situation described by the Lagrangian

$$\mathcal{L}(\phi) = \frac{1}{2}\partial_\mu\phi\,\partial^\mu\phi - V(\phi) \tag{2.5.1}$$

where ϕ is some multiplet of spinless fields and $V(\phi)$ and thus $\mathcal{L}(\phi)$ is invariant under some symmetry group

$$\delta\phi = i\epsilon_a T^a\phi \tag{2.5.2}$$

where the T^a are imaginary antisymmetric matrices (because ϕ are Hermitian).

As in the previous section, we want to perturb around a minimum of the potential $V(\phi)$. We expect the ϕ field to have a VEV, $\langle\phi\rangle = \lambda$, which minimizes V. To simplify the analysis, we define

$$V_{j_1\cdots j_n}(\phi) = \frac{\partial^n}{\partial\phi_{j_1}\cdots\partial\phi_{j_n}}V(\phi) \tag{2.5.3}$$

Then we can write the condition that λ be an extremum of $V(\phi)$ as

$$V_j(\lambda) = 0 \tag{2.5.4}$$

Since λ is a minimum, V must also satisfy

$$V_{jk}(\lambda) \geq 0 \tag{2.5.5}$$

The second derivative matrix $V_{jk}(\lambda)$ is the meson mass-squared matrix. We can see this by expanding $V(\phi)$ in a Taylor series in the shifted fields $\phi' = \phi - \lambda$ and noting that the mass term is $\frac{1}{2}V_{jk}(\lambda)\phi'_j\phi'_k$. Thus, (2.5.5) assures us that there are no tachyons in the free theory about which we are perturbing.

Now comes the interesting part, the behavior of the VEV λ under the transformations (2.5.2). There are two cases. If

$$T_a\lambda = 0 \tag{2.5.6}$$

for all a, the symmetry is not broken. This is certainly what happens if $\lambda = 0$. But (2.5.6) is the more general statement that the vacuum doesn't carry the charge T_a, so the charge cannot disappear into the vacuum. But it is also possible that

$$T_a\lambda \neq 0 \quad \text{for some } a \tag{2.5.7}$$

Then the charge T_a can disappear into the vacuum even though the associated current is conserved. This is spontaneous symmetry breaking.

Often there are some generators of the original symmetry that are spontaneously broken while others are not. The set of generators satisfying (2.5.6) is closed under commutation (because $T_a\lambda = 0$ and $T_b\lambda \Rightarrow [T_a\,,\,T_b]\,\lambda = 0$) and generates the unbroken subgroup of the original symmetry group.

Now let us return to the mass matrix. Because V is invariant under (2.5.2), we can write

$$V(\phi + \delta\phi) - V(\phi) = iV_k(\phi)\epsilon_a(T^a)_{kl}\phi_l = 0 \tag{2.5.8}$$

If we differentiate with respect to ϕ_j, we get (since ϵ^a are arbitrary)

$$V_{jk}(\phi)(T^a)_{kl}\phi_l + V_k(\phi)(T^a)_{kj} = 0 \tag{2.5.9}$$

Setting $\phi = \lambda$ in (2.5.9), we find that the second term drops out because of (2.5.4), and we obtain

$$V_{jk}(\lambda)(T^a)_{kl}\lambda_l = 0 \tag{2.5.10}$$

But $V_{jk}(\lambda)$ is the mass-squared matrix M^2_{jk} for the spinless fields, so we can rewrite (2.5.10) in matrix form as

$$M^2T^a\lambda = 0 \tag{2.5.11}$$

For T_a in the unbroken subgroup, (2.5.11) is trivially satisfied. But if $T^a\lambda \neq 0$, (2.5.11) requires that $T^a\lambda$ is an eigenvector of M^2 with zero eigenvalue. It corresponds to a massless boson field, given by

$$\phi^T T^a \lambda \tag{2.5.12}$$

This is called a Goldstone boson after J. Goldstone, who first established this connection between spontaneously broken continuous symmetries and massless particles.

The existence of the massless particle can be understood qualitatively as follows. If the symmetry is spontaneously broken, we know that our vacuum is part of a continuous set of degenerate vacua that can be rotated into one another by the symmetry. Physically, we cannot really get from one vacuum to another because to do so would require transformation of our local fields everywhere in space-time. But we can look at states that differ from our vacuum by such a rotation in a finite region and then go smoothly back to our vacuum outside. The point is that the energy of such a state can be made arbitrarily close to the energy of our vacuum state by making the region larger and the transition smoother. But if there are states in the theory with energy arbitrarily close to the energy of the vacuum state, then there must be massless particles in the theory. These are Goldstone bosons. Their masslessness is the translation into local field theory of the global degeneracy of the vacuum.

2.6 The σ-Model

Before discussing the spontaneous breakdown of symmetry in gauge theories, we will work out in detail one example of global symmetry breakdown in the strong interactions that will be useful when we discuss hadrons — and will turn out to have a curious connection with the weak interactions as well. The example is the σ-model of Gell-Mann and Levy (*Nuovo Cimento* **16**:705-713, 1960), a toy model of nuclear forces and, in particular, of the π-nucleon coupling.

Let ψ be an isospin-doublet field representing the nucleons, P and N,

$$\psi = \begin{pmatrix} P \\ N \end{pmatrix} \tag{2.6.1}$$

The theory should certainly be invariant under global rotations, with $T_a = \tau_a/2$

$$\delta\psi = i\epsilon^a T_a \psi \tag{2.6.2}$$

However, as we have seen, the kinetic-energy term for massless fermions is automatically invariant under the larger group of symmetries, $SU(2) \times SU(2)$:

$$\begin{aligned} \delta\psi_L &= i\epsilon_L^a T_a \psi_L \\ \delta\psi_R &= i\epsilon_R^a T_a \psi_R \end{aligned} \tag{2.6.3}$$

These are the chiral symmetries. (2.6.3) can be rewritten in terms of the infinitesimal parameters

$$\begin{aligned} \epsilon^a &= (\epsilon_R^a + \epsilon_L^a)/2 \\ \epsilon_5^a &= (\epsilon_R^a - \epsilon_L^a)/2 \end{aligned}, \tag{2.6.4}$$

in the following form

$$\delta\psi = i(\epsilon^a - \gamma_5 \epsilon_5^a) T_a \psi \tag{2.6.5}$$

If $\epsilon_5^a = 0$, this is a pure isospin rotation. If $\epsilon^a = 0$, it's a pure chiral rotation.

This would all seem to be academic, since a nucleon mass term $\overline{\psi}\psi$ breaks the chiral symmetry and leaves only isospin. But Gell-Mann and Levy found that they could build a Lagrangian with chiral symmetry and a nucleon mass if the chiral symmetry was spontaneously broken. In the process, the pion is interpreted as a Goldstone boson.

The Lagrangian involves a 2×2 matrix of spinless fields Σ that transforms as follows under the chiral symmetries:

$$\delta\Sigma = i\epsilon_L^a T_a \Sigma - i\Sigma\epsilon_R^a T_a \tag{2.6.6}$$

Then the Lagrangian has the following form:

$$\mathcal{L} = i\overline{\psi}\,\partial\!\!\!/\,\psi - g\overline{\psi_L}\Sigma\psi_R - g\overline{\psi_R}\Sigma^\dagger\psi_L + \mathcal{L}(\Sigma) \tag{2.6.7}$$

The invariance of the Yukawa couplings may be more transparent in terms of finite transformations. The infinitesimal transformations (2.6.3) and (2.6.6) can be integrated to obtain

$$\psi_L \to L\psi_L, \quad \psi_R \to R\psi_R \tag{2.6.8}$$

$$\Sigma \to L\Sigma R^\dagger \tag{2.6.9}$$

where L and R are independent 2×2 unitary matrices with determinant 1,

$$L = \exp\left(il^a \tau_\alpha\right), \quad R = exp\left(ir^a \tau_a\right) \tag{2.6.10}$$

with l^a and r^a arbitrary real 3-vectors.

The most general 2×2 matrix would have eight real components. The Σ field is constrained to depend only on four real fields, as follows:

$$\Sigma = \sigma + i\tau_a \pi_a \tag{2.6.11}$$

It is not obvious (at least to me) that this form is preserved by the transformations (2.6.6).[1] But it is true, and you can work out by explicit calculation the transformations of the σ and π_a fields:

$$\begin{aligned} \delta\sigma &= \epsilon_5^a \pi_a \\ \delta\pi_a &= -\epsilon_{abc}\epsilon^b \pi_c - \epsilon_5^a \sigma \end{aligned} \tag{2.6.12}$$

Another way to see this is to note that $\Sigma^\dagger \Sigma = \Sigma\Sigma^\dagger = (\sigma^2 + \vec{\pi}^2)\, I$ and $\det \Sigma = \sigma^2 + \vec{\pi}^2$. Thus Σ is $\sqrt{\sigma^2 + \vec{\pi}^2}$ times a unitary unimodular matrix. Obviously, if we multiply Σ on the left or on the right by a unitary unimodular matrix, the result is still of the form $\sqrt{\sigma^2 + \vec{\pi}^2}$ times a unitary unimodular matrix.

Inserting (2.6.11) into $\mathcal{L}$, we can write the Yukawa couplings as

$$-\,g\sigma\overline{\psi}\psi + ig\pi_a\overline{\psi}\gamma_5\tau_a\psi \tag{2.6.13}$$

From (2.6.13), you can see that the π_a fields have the right form to describe the π's. The coupling g is the πNN coupling $g_{\pi NN}$.

We still don't have a nucleon mass term, but it is clear from (2.6.13) that if we can give σ a VEV, we will be in good shape. To this end, we must ask how to build $\mathcal{L}(\Sigma)$ invariant under $SU(2) \times SU(2)$. It is clear from (2.6.9) that

$$\frac{1}{2}\mathrm{tr}\left(\Sigma^\dagger\Sigma\right) = \sigma^2 + \pi_a^2 \tag{2.6.14}$$

is invariant. In fact, the most general invariant (without derivatives) is just a function of $\sigma^2 + \pi_a^2$. This can be seen by noting that more complicated traces just give powers of $\sigma^2 + \pi_a^2$ because

$$\Sigma^\dagger\Sigma = \left(\sigma^2 + \pi_a^2\right) \tag{2.6.15}$$

and is proportional to the identity in the 2×2 space. Alternatively, we can recognize (2.6.12) as the transformation law of a 4-vector in four-dimensional Euclidean space. The invariant (2.6.14) is just the length of the vector, the only independent variant.

Without further introduction, we can write down the invariant $\mathcal{L}$,

$$\mathcal{L}(\Sigma) = \frac{1}{2}\partial^\mu\sigma\partial_\mu\sigma + \frac{1}{2}\partial^\mu\pi_a\partial_\mu\pi_a - V\left(\sigma^2 + \pi_a^2\right) \tag{2.6.16}$$

To get a VEV, we take V to be

$$V(\sigma^2 + \pi^2) = \frac{\lambda}{4}\left[\left(\sigma^2 + \pi_a^2\right)^2 - F_\pi^2\right]^2 \tag{2.6.17}$$

[1]These objects are actually related to interesting things called quaternions.

where λ is a dimensionless constant and F_π has dimensions of mass (indeed, it is the only mass scale in the theory so far). Then $\sigma = \pi_a = 0$ is not a minimum. V is obviously minimized for

$$\sigma^2 + \pi_a^2 = F_\pi^2 \tag{2.6.18}$$

Now we can use our freedom to make $SU(2) \times SU(2)$ transformations to rotate any VEV into the σ direction so that without any loss of generality we can assume

$$\langle\sigma\rangle = F_\pi, \quad \langle\pi_a\rangle = 0 \tag{2.6.19}$$

and perturb around that vacuum.

Thus, we define the shifted field

$$\sigma' = \sigma - F_\pi, \quad \langle\sigma'\rangle = 0 \tag{2.6.20}$$

in terms of which $\mathcal{L}$ is

$$\begin{aligned} \mathcal{L} \; &= i\overline{\psi}\not{\partial}\psi - gF_\pi\overline{\psi}\psi - g\sigma'\overline{\psi}\psi \\ &+ig\pi_a\overline{\psi}\tau_a\gamma_5\psi + \tfrac{1}{2}\partial^\mu\sigma'\partial_\mu\sigma' + \tfrac{1}{2}\partial^\mu\pi_a\partial^\mu\pi_a \; \cdot \\ &-\tfrac{\lambda}{4}\left(\sigma'^2 + \pi_a^2 + 2F_\pi\sigma'\right)^2 \end{aligned} \tag{2.6.21}$$

This describes nucleons with mass gF_π coupled to the scalar σ' field and massless pseudoscalar π_a's.

Why did Gell-Mann and Levy think (2.6.21) had anything to do with the world? For one thing, the physical pion is very light compared to other hadrons. For example, $m_\pi^2/m_N^2 \simeq 1/50$. Perhaps a theory in which it is massless is not such a bad approximation. But there was another reason. The parameter g can clearly be measured in π-nucleon interactions. It turns out that F_π can also be measured. The reason is that, as we shall discuss in enormous detail later, it determines the rate at which $\pi^\pm$ decay through the weak interactions. The crucial fact is that the axial vector current, the current associated with ϵ_5^a transformations, has the form

$$j_{5a}^\mu = -\left(\partial^\mu\pi_a\right)\sigma + \left(\partial^\mu\sigma\right)\pi_a - \overline{\psi}\gamma^\mu\gamma_5\tau_a\psi \tag{2.6.22}$$

which in terms of shifted fields has a piece proportional to $\partial^\mu\pi_a$:

$$j_{5a}^\mu = -F_\pi\partial^\mu\pi_a + \cdots \tag{2.6.23}$$

The other terms are all bilinear in the fields. The point is that this current has a nonzero matrix element between the vacuum and a one-pion state

$$\langle 0|j_{5a}^\mu|\pi_b\rangle = iF_\pi p^\mu\delta_{ab} \tag{2.6.24}$$

where p^μ is the pion momentum. This is odd. A normal current, like the charge with which it is associated, just moves you around within multiplets. (2.6.24) is a sign of spontaneous symmetry breaking. At any rate, the decay $\pi^+ \to \mu^+\nu_\mu$ is proportional to F_π^2, and so F_π can be measured. Then the nucleon mass can be predicted according to

$$m_N = g_{\pi NN}F_\pi \tag{2.6.25}$$

This relation is called the Goldberger-Treiman relation, and it works fairly well (actually (2.6.25) us a special case of the general Goldberger-Treiman relation that works even better).

The pion is a Goldstone boson in this model. This is probably obvious, but we can use the formal machinery of Section 2.5 to see it directly. The Goldstone-boson directions are defined by $T_a\lambda$ where λ is the VEV. In this theory, the isospin generators annihilate the vacuum, so isospin is not spontaneously broken and there is no scalar Goldstone boson. But the chiral transformations rotate the VEV into the π directions ($\sigma\pi_a = \cdots + \epsilon_5^a F_\pi$ from (2.6.12)), and thus the chiral symmetry is broken and the π's are Goldstone bosons.

We will discuss explicit chiral symmetry breaking later, but now notice that we can incorporate a pion mass by adding to V the term

$$\frac{m_\pi^2}{2}\left(\sigma^2 + \pi^2 - 2F_\pi\sigma + F_\pi^2\right) \tag{2.6.26}$$

which is not invariant because of the linear term $m_\pi^2 F_\pi \sigma$.

2.7 The Higgs Mechanism

We now want to apply the idea of spontaneous symmetry breaking to the $SU(2)\times U(1)$ model of leptons. It seems clear that we can get an electron mass in much the same way we get nucleon masses in the σ-model. It is probably not obvious that the $W^\pm$ and Z will get mass. But wait and see. To give mass to the electron, we need an $SU(2)\times U(1)$ multiplet of spinless fields that can couple the ψ_L to the e_R^- in a Yukawa coupling,

$$-f\,\overline{e_R^-}\,\phi^\dagger\psi_L + \text{h.c.} \tag{2.7.1}$$

If the field ϕ transforms under $SU(2)\times U(1)$ with charges

$$\vec{T}\phi = \frac{\vec{\tau}}{2}\phi, \quad S\phi = \frac{1}{2}\phi\,, \tag{2.7.2}$$

then (2.2.6) and (2.2.10) with (2.7.2) imply that (2.7.1) is invariant under the $SU(2)$ transformation $\delta = i\epsilon_a T^a + i\epsilon S$. Evidently, ϕ must be a doublet field

$$\begin{pmatrix}\phi^+\\ \phi^0\end{pmatrix} = \phi \tag{2.7.3}$$

where the superscripts are the Q values according to $Q = T_3 + S$. Note that

$$\delta\phi^\dagger = -i\epsilon_a\phi^\dagger T_a - \frac{i}{2}\epsilon\phi^\dagger \tag{2.7.4}$$

so that

$$\delta\left(\phi^\dagger\psi_L\right) = -i\epsilon\left(\phi^\dagger\psi_L\right) \tag{2.7.5}$$

$\phi^\dagger\psi_L$ is an $SU(2)$ singlet with $S + -1$, just like e_R^-, so (2.7.1) is invariant. Now when ϕ^0 has a nonzero VEV, (2.7.2) has a piece that looks like an electron mass.

Now that we have the $SU(2)\times U(1)$ properties of ϕ determined, we know the couplings of the gauge particles to the ϕ — they are determined by the covariant derivatives:

$$\mathcal{L}_{KE}(\phi) = (D_\mu\phi)^\dagger(D^\mu\phi) \tag{2.7.6}$$

We can do everything in this notation, but it is slightly more convenient to go over to a notation in which the scalar fields are self-adjoint. We can easily do this by rewriting the complex ϕ^+ and ϕ^0 fields in terms of their real and imaginary parts:

$$\begin{pmatrix}\phi^+\\ \phi^0\end{pmatrix} = \begin{pmatrix}(\phi_3 + i\phi_4)/\sqrt{2}\\ (\phi_1 + i\phi_2)/\sqrt{2}\end{pmatrix} \tag{2.7.7}$$

The conventional $\sqrt{2}$ is intended to ensure that the fields are normalized in the same way. At any rate, we can arrange the real fields ϕ_j in a real 4-vector and ask how the generators $\vec{T}$ and S translate into this notation. Thus, we write

$$\Phi = \begin{pmatrix} \phi_3 \\ \phi_4 \\ \phi_1 \\ \phi_2 \end{pmatrix} \tag{2.7.8}$$

The space of the Φ field has an obvious tensor-product structure. It is a tensor product of the two-dimensional space of the original ϕ, on which the τ's act, and a two-dimensional space corresponding to the real and imaginary parts of the components of ϕ, on which we can define an independent set of Pauli matrices $\vec{\sigma}$. The 15 traceless Hermitian matrices acting on ϕ can be written as

$$\sigma_j,\ \tau_j,\ \text{and}\ \sigma_j\tau_k \quad (\text{where } j,\ k = 1 \text{ to } 3) \tag{2.7.9}$$

In this notation, it is easy to write down the generators $\vec{T}$ and S. The procedure is simple and quite general: Leave antisymmetric matrices in ϕ space unchanged, but multiply symmetric matrices by $-\sigma_2$ to make them antisymmetric. Thus

$$\begin{aligned} T_1\Phi &= -\tfrac{1}{2}\tau_1\sigma_2\Phi \\ T_2\Phi &= \tfrac{1}{2}\tau_2\Phi \\ T_3\Phi &= -\tfrac{1}{2}\tau_3\sigma_2\Phi \\ S\Phi &= -\tfrac{1}{2}\sigma_2\Phi \end{aligned} \tag{2.7.10}$$

You can easily check that (2.7.10) gives the same transformation laws as (2.7.2) and (2.7.7). The advantage of this notation is really marginal. For example, it makes it slightly easier to identify the Goldstone bosons. But since it is so easy, we might as well use it.

In terms of Φ, the kinetic-energy term becomes

$$\begin{aligned} \mathcal{L}_{KE}(\Phi) &= \tfrac{1}{2}D^\mu\Phi^T D_\mu\Phi \\ &= \tfrac{1}{2}\left[\partial^\mu\Phi^T - i\Phi^T\left(\tfrac{e}{\sin\theta}\vec{T}\cdot\vec{W}_\mu + \tfrac{e}{\cos\theta}SX_\mu\right)\right] \\ &\quad \left[\partial_\mu + i\left(\tfrac{e}{\sin\theta}\vec{T}\cdot\vec{W}_\mu + \tfrac{e}{\cos\theta}SX_\mu\right)\right]\Phi \end{aligned} \tag{2.7.11}$$

The point of all this is that (2.7.11) contains terms like $\Phi^T W_1^\mu W_{1\mu}\Phi$. If Φ has a VEV, this looks like a W_1 mass. Perhaps we will be able to give mass to the $W^\pm$ and Z. Of course, we hope to do it without giving mass to the photon, and we still have the Goldstone bosons to worry about. But let's go on.

The most general $SU(2)\times U(1)$ invariant potential depends only on the combination $\phi^\dagger\phi$. In particular, if it has the form

$$V(\phi) = \frac{\lambda}{2}\left(\phi^\dagger\phi - v^2/2\right)^2 \tag{2.7.12}$$

Then ϕ will develop a VEV.

We want the VEV of ϕ^0 to be real and positive to that (2.7.1) gives an electron mass term that is real and positive. This is purely conventional, of course. We can make an $SU(2) \times U(1)$ transformation to make any VEV of ϕ^+ and ϕ^0 have the form

$$\langle \phi^+ \rangle = 0, \quad \langle \phi^0 \rangle = v/\sqrt{2} \tag{2.7.13}$$

for real v. Let's prove that. Suppose

$$\langle \phi \rangle = \begin{pmatrix} a \\ b \end{pmatrix} \tag{2.7.14}$$

By making a transformation of the form

$$\phi \to \exp(2i\theta_3 T_3)\theta$$

we get

$$a \to e^{i\theta_3} a, \quad b \to e^{-1\theta_3} b \tag{2.7.15}$$

By a suitable choice of θ_3 we can make a and b have the same phase. Then

$$\phi \to \exp(2i\phi_2 T_2)\phi \tag{2.7.16}$$

is an orthogonal transformation by which we can rotate the $(a,\ b)$ vector into the ϕ^0 direction. Then by another (2.7.15) transformation we can make the VEV real and positive. Thus, all we have done is to choose a convenient form for the VEV so that our original labeling of the fermion fields is consistent.

In any case, (2.7.13) is equivalent to

$$\langle \Phi \rangle = \begin{pmatrix} 0 \\ 0 \\ v \\ 0 \end{pmatrix} \tag{2.7.17}$$

Now we want to rewrite (2.7.13) in terms of the VEV, λ, and the shifted scalar field:

$$\Phi' = \Phi - \lambda \tag{2.7.18}$$

Of course, $\mathcal{L}_{KE}(\Phi)$ contains the kinetic energy term for the Φ' fields. But for the moment we will concentrate on the terms that are quadratic in the fields. There are two types:

$$\frac{1}{2}\lambda^T \left[\frac{e}{\sin\theta} \vec{T} \cdot \vec{W}^\mu + \frac{e}{\cos\theta} S X^\mu \right] \cdot \left[\frac{e}{\sin\theta} \vec{T} \cdot W_\mu + \frac{e}{\cos\theta} S X_\mu \right] \lambda \tag{2.7.19}$$

and

$$i\partial^\mu \Phi'^{\,T} \left[\frac{e}{\sin\theta} \vec{T} \cdot \vec{W}_\mu + \frac{e}{\cos\theta} S X_\mu \right] \lambda \tag{2.7.20}$$

The (2.7.19) are the W and Z mass terms we want, as we will soon see. But (2.7.20) looks dangerous. It describes some sort of mixing between the Goldstone bosons and the gauge fields. The Goldstone-boson fields are

$$\Phi'^{\,T} \vec{T} \lambda = \Phi^T \vec{T} \lambda \tag{2.7.21}$$

The field $\Phi'^{\,T} S \lambda$ is not independent, since $T_3 \lambda = -S\lambda$.

But we have not yet used all our freedom to make local $SU(2) \times U(1)$ transformations. In fact, we can choose a gauge, called the unitary gauge, in which the Goldstone-boson fields just vanish, so we can throw (2.7.20) away. To see this, we need to return to the unbroken theory. We saw in (2.7.14-17) that we use the global $SU(2) \times U(1)$ symmetry to take an arbitrary VEV and rotate it into the ϕ_3 direction. But since we have the freedom to make different $SU(2) \times U(1)$ transformations at each point in space-time, we can take an arbitrary *field* $\Phi(x)$ and rotate it into the ϕ_1 direction! So that after these rotations

$$\phi_2(x) = \phi_3(x) = \phi_4(x) = 0 \tag{2.7.22}$$

Of course, we now have no further freedom to rotate Φ without disturbing (2.7.22). In other words, we have chosen a gauge. Comparing (2.7.21) and (2.7.22), we can see explicitly that the Goldstone-boson fields vanish in this gauge. The VEV is a VEV of the single remaining field ϕ_3.

Having disposed of (2.7.20), we can now return to (2.7.19) and show that it is a mass term for the $W^\pm$ and the Z. Note first that we can separate the problem into a neutral-gauge boson mass matrix and a separate-charged boson matrix. Electromagnetic charge conservation prevents mixing between the two. The neutral sector is more complicated than the charged sector, so we will do it first. We saw in Section 2.2 that we could rewrite the covariant derivative in terms of A^μ and Z fields as follows:

$$\frac{e}{\sin\theta} T_3 W_3^\mu + \frac{e}{\cos\theta} S X^\mu = eQA^\mu + \frac{e}{\sin\theta\cos\theta}\left(T_3 - \sin^2\theta\, Q\right) Z^\mu \tag{2.7.23}$$

Now we can see why the photon field doesn't get a mass — because

$$Q\lambda = 0 \tag{2.7.24}$$

so that A^μ doesn't appear in (2.7.19) and (2.7.20). This, of course, is the way it had to work out. (2.7.24) is the statement that the electromagnetic gauge invariance is not broken by the vacuum. Thus, it must imply that the photon remains massless, and it does. (2.7.24) also shows that the $\sin^2\theta Q$ term in the Z^μ coupling is irrelevant to the mass. Putting (2.7.23)-(2.7.24) into (2.7.19)-(2.7.20), we get

$$\frac{1}{2}\frac{e^2}{\sin^2\theta\cos^2\theta} Z^\mu Z_\mu \lambda^T T_3^2 \lambda = \frac{1}{2}\left[\frac{ev}{2\sin\theta\cos\theta}\right]^2 Z^\mu Z_\mu \tag{2.7.25}$$

Thus

$$M_Z = \frac{ev}{2\sin\theta\cos\theta} \tag{2.7.26}$$

An even simpler analysis for the charged fields gives

$$\frac{1}{2}\left[\frac{ev}{2\sin\theta}\right]^2 (W_1^\mu W_{1\mu} + W_2^\mu W_{2\mu}) \tag{2.7.27}$$

Thus

$$M_W = \frac{ev}{2\sin\theta} = M_Z \cos\theta \tag{2.7.28}$$

The gauge defined by (2.7.22) is called "unitary" or "unitarity" gauge because all fields that appear correspond to physical particles that appear in the S-matrix. In general, the unitary-gauge requirement is that the Goldstone-boson fields vanish,

$$\Phi^T T_a \lambda = 0 \tag{2.7.29}$$

In this gauge, renormalizability is not obvious. But as 't Hooft showed, the theory is renormalizable. We will discuss other gauges and the question of radiative corrections in Chapter 8.

2.8 Neutral Currents

As you will show in Problem 2-1, at low-momentum transfers, $W^\pm$ exchange gives rise to the effective four-fermion interaction (2.3.1) with

$$G_F = \frac{\sqrt{2}\,e^2}{8M_W^2 \sin^2\theta} = \frac{1}{\sqrt{2}v^2} \tag{2.8.1}$$

This is just one term of the general charged-current interaction that has the following form:

$$\frac{G_F}{\sqrt{2}}\left(j_1^\alpha j_{1\alpha} + j_2^\alpha j_{2\alpha}\right) \tag{2.8.2}$$

where the currents are

$$j_a^\alpha = \sum_\psi \overline{\psi_L} T_a \gamma^\alpha (1+\gamma_5)\psi_l \tag{2.8.3}$$

where T_a are the $SU(2)$ generators. So far, we have seen only the leptonic contribution to (2.8.3), but as we will see later, there are others.

The coupling of the Z produces a neutral-current interaction analogous to (2.8.4). In the same notation in which the $W^\pm$ coupling is

$$\frac{e}{\sin\theta} T_a \tag{2.8.4}$$

the Z^μ coupling is

$$\frac{e}{\sin\theta\cos\theta}\left(T_3 - \sin^2\theta\, Q\right) \tag{2.8.5}$$

However, since the Z is heavier that the W by a factor of $1/\cos\theta$, the extra factor of $\cos\theta$ in (2.8.5) cancels from the effective four-fermion interaction, which has the form

$$\frac{G_F}{\sqrt{2}}\left(j_3^\alpha - 2\sin^2\theta\, j_{EM}^\alpha\right)\cdot\left(j_{3\alpha} - 2\sin^2\theta\, j_{EM\alpha}\right)\ \left(\text{where } j_{EM}^\alpha = \overline{\psi}\gamma^\alpha Q\psi\right) \tag{2.8.6}$$

The normalization of (2.8.6) is easy to remember because the j^2 term has the same strength as the j_1^2 and j_2^2 terms in (2.8.2).

The fact that the coefficients of j_1^2 and j_2^2, and j_3^2 are the same is really much more than mnemonic. It is symptomatic of an important feature of the theory. In the limit in which the g' coupling is turned off, the theory has a larger symmetry. Consider the substitutions

$$\phi_1 = \sigma, \quad \phi_2 = -\pi_3, \quad \phi_3 = \pi_2, \quad \phi_4 = \pi_1 \tag{2.8.7}$$

Comparing (2.7.2) and (2.6.12), we see that our ϕ doublet can be identified with the Σ field of the $SU(2)_L \times SU(2)_R$ σ-model, and the weak $SU(2)$ is just $SU(2)_L$. If there were no g' coupling and no Yukawa couplings, the theory would have a gauged $SU(2)_L$ symmetry and a global $SU(2)_R$ symmetry. When ϕ_1 ($=\sigma$) gets a VEV, the $SU(2)_L \times SU(2)_R$ symmetry is broken down to the diagonal $SU(2)$, which remains as an unbroken global symmetry of the theory. This $SU(2)$ is sometimes called the "custodial" $SU(2)$ symmetry. It requires the neutral currents to have the form (2.8.6) because they must be part of an $SU(2)$ singlet with (2.8.2) when $\sin^2\theta = 0$.

The parameter G_F, the Fermi constant, can be fixed by a measurement of the μ decay rate (see Problem 2-4). After appropriate radiative corrections, we find

$$G_F = 1.166 \times 10^{-5}\ \text{GeV}^{-2} \tag{2.8.8}$$

(2.8.6) has some interesting physics associated with it. Consider first the general question of weak cross sections. If some scattering process takes place purely through the weak interactions, the amplitude is proportional to G_F, and so the cross section is proportional to G_F^2. Thus, σ is proportional to G_F^2 times a function of s and the fermion masses, where s is the square of the center of mass energy. Furthermore, there is no way for one of the fermion masses to get into the denominator. Thus, for s large compared to fermion masses, we must have

$$\sigma \simeq G_F^2 s \tag{2.8.9}$$

on dimensional grounds.

(2.8.9) has important consequences for weak-scattering experiments. Since the only convenient lepton fixed target is the e^-, it is hard to get a large s in fixed-target experiments because

$$s \simeq 2Em_e \tag{2.8.10}$$

where E is the beam energy. Thus, to see processes such as

$$\nu_\mu + e^- \to \nu_\mu + e^- \quad \text{or } \mu^- + \nu_e \tag{2.8.11}$$

$$\overline{\nu_\mu} + e^- \to \overline{\nu_\mu} + e^- \tag{2.8.12}$$

$$\overline{\nu_\mu} + e^- \to \overline{\nu_e} + e^- \tag{2.8.13}$$

is not easy. The first two have been seen in high-energy neutrino-scattering experiments where the neutrinos are primarily ν_μ and $\overline{\nu_\mu}$ from $\pi^\pm$ decay. (2.8.13) has been seen in low-energy antineutrino beams from reactors where enormous neutrino fluxes are available. The advantage of these experiments as tests of the model are that they only involve leptons — no strong interactions. But the experiments are so hard that this is not the easiest way to get really detailed tests of the $SU(2) \times U(1)$ theory. It works, but there is also good evidence from neutrino-nucleon scattering, which we will discuss in some detail later. The data are better for neutrino-nucleon scattering. The cross sections are larger because s is larger for a given beam energy by m_{Nucleon}/m_e — a large factor. We will study this in Chapter 7.

2.9 $e^+e^- \to \mu^+\mu^-$

The situation is more complicated in e^+e^- annihilation. At the low center of mass energies, the process $e^+e^- \to \mu^+\mu^-$ is dominated by electromagnetic interactions, the s-channel exchange of a virtual proton shown in Figure 2-4.

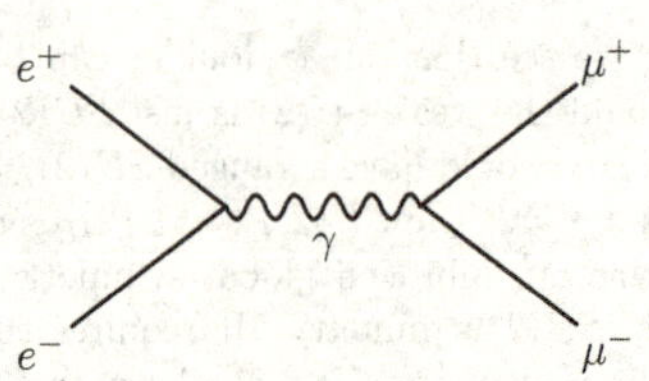

Figure 2-4:

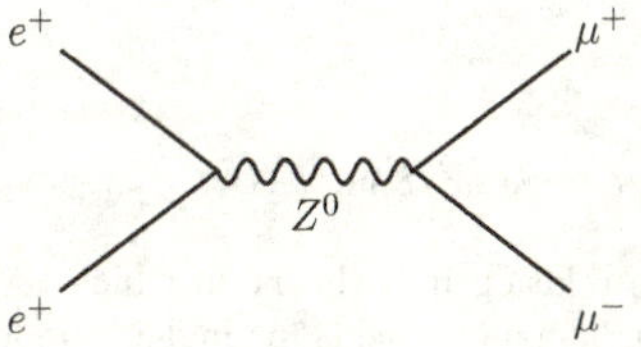

Figure 2-5:

This produces a total cross section for the process that falls with $1/s$ ($s = 4E^2$, where E is the center of mass energy of one of the particles). The cross section (at energies large compared to the particle masses) is

$$\sigma_{e^+e^- \to \mu^+\mu^-} = \frac{4\pi\alpha^2}{3s} \tag{2.9.1}$$

But the neutral-current weak interactions also contribute, as seen in Figure 2-5. Indeed, when the total center of mass energy approaches the Z^0 mass, this becomes the dominant contribution to the amplitude. Eventually, this process should be one of the best sources of information about the Z^0.

In the 70's and early 80's, the largest active e^+e^- machines were the PETRA machine in Hamburg and the PEP machine in Palo Alto, with a total center of mass energy between 30 and 40 GeV. This was not enough to produce the Z^0 resonance. The largest effect of what occurs in Figure 2-5 at these energies is the interference between the processes in Figures 2-4 and 2-5. There are two types of contribution to the interference in the total cross section. Since the Z^0 has both vector (V) and axial vector (A) couplings, the process in Figure 2-5 has pieces of the form VV, AA, and AV. But the photon has only V couplings, so (2.8.1) has the form VV. Thus the interference has one of the following forms:

$$\text{VV VV} \tag{2.9.2}$$

$$\text{VV AA} \tag{2.9.3}$$

$$\text{VV VA} \tag{2.9.4}$$

(2.9.4) does not contribute at all to the total cross section because it is a pseudoscalar. It does contribute to various parity violating effects, such as helicity dependence.

(2.9.2) shares an interesting property with the purely electromagnetic contribution in that it looks the same if we interchange the μ^+ and the μ^- in the final state. If this were the only contribution, as many μ^+ would scatter forward in the direction of the incoming e^+ as backward, in the opposite direction (see Problem 2-7).

(2.9.3), on the other hand, changes sign under the interchange of μ^+ and μ^- because the V and A currents have opposite-charge conjugation. Thus, this term contributes to a front-back asymmetry in the cross section (see Problem 2-7). The front-back asymmetry has been measured and agrees with the $SU(2) \times U(1)$ model, not only for $e^+e^- \to \mu^+\mu^-$, but also for $e^+e^- \to \tau^+\tau^-$. This was exciting because it was the first data we had on the τ neutral current.

Today, the largest e^+e^- machine is LEP at CERN which runs at a center of mass energy of about M_Z and today has produced millions of Zs. The SLC, the novel single pass collider at SLAC also produces Zs in e^+e^- annihilation. In this machine, the electrons can be polarized, so the interactions of the left-handed and right-handed electrons can be separately studied.

Problems

2-1. Derive the relation $M_W^2 = \sqrt{2}e^2/8\sin^2\theta\, G_F$.

2-2. Derive the analogous relation in a theory in which the left- handed electron and muon fields are in triplets (with all right-handed fields in singlets) under $SU(2)\times U(1)$ as follows:

$$\psi_{eL} = \begin{pmatrix} E_L^+ \\ \nu_{eL} \\ e_L^- \end{pmatrix} \qquad \psi_{\mu L} = \begin{pmatrix} M_L^+ \\ \nu_{\mu L} \\ \mu_L^- \end{pmatrix}$$

where E^+ and M^+ are heavy (unobserved) lepton fields.

The $SU(2)$ generators of the triplet representation are

$$T_1 = \frac{1}{\sqrt{2}}\begin{pmatrix} 0 & 1 & 0 \\ 1 & 0 & 1 \\ 0 & 1 & 0 \end{pmatrix} \qquad T_2 = \frac{i}{\sqrt{2}}\begin{pmatrix} 0 & -i & 0 \\ i & 0 & -i \\ 0 & i & 0 \end{pmatrix}$$

$$T_3 = \begin{pmatrix} 1 & 0 & 0 \\ 0 & 0 & 0 \\ 0 & 0 & -1 \end{pmatrix}$$

Note that $[T_a, T_b] = i\epsilon_{abc}T_c$

2-3. Find the constants C_j in the following relations:

(a) $[\gamma^\mu P_\pm]_{ij}[\gamma_\mu P_\pm]_{kl} = C_1[\gamma^\mu P_\pm]_{il}[\gamma_\mu P_\pm]_{kj}$
(b) $[\gamma^\mu P_\pm]_{ij}[\gamma_\mu P_\mp]_{kl} = C_2[P_\mp]_{il}[P_\pm]_{kj}$
(b′) Note that $[\sigma^{\mu\nu}P_\mp]_{il}[\sigma_{\mu\nu}P_\pm]_{kj}$
(c) $[P_\pm]_{ij}[P_\pm]_{kl} = C_3[P_\pm]_{il}[P_\pm]_{kj} + C_4[\sigma^{\mu\nu}P_\pm]_{il}[\sigma_{\mu\nu}P_\pm]_{kj}$
(d) $[\sigma^{\mu\nu}P_\pm]_{ij}[\sigma_{\mu\nu}P_\pm]_{kl} = C_3[P_\pm]_{il}[P_\pm]_{kj} + C_4[\sigma^{\mu\nu}P_\pm]_{il}[\sigma_{\mu\nu}P_\pm]_{kj}$

These six numbers are all there is to Fierz transformations.

Hint: For (a) and (b), multiply by γ^ν_{lk} and sum over l and k. For (c) and (d), it's easiest to avoid taking traces of $\sigma^{\mu\nu}$, so multiply by δ_{lk} and sum over l and k to get one equation. Multiply by δ_{jk} and sum over j and k to get another. To prove (b′), multiply by γ^λ_{lk} and sum over l and k.

2-4. Calculate the μ decay rate assuming $M_W \gg m_\mu m_e$. Note that in this limit, the rate depends only on G_F and m_e.

2-5. In the $SU(2)\times U(1)$ theory, both $W^\pm$ and Z^0 exchange contribute to the elastic scattering process, $\overline{\nu_e}e^- \to \overline{\nu_e}e^-$. For momentum transfers very small compared to M_W, use the results of Problem 2-3 to write the effective interaction Hamiltonian for this process in the form:

$$\frac{G_F}{\sqrt{2}}\,\overline{\nu_e}\gamma_\mu(1+\gamma_5)\nu_e\,\overline{e^-}\gamma^\mu(G_V + G_A\gamma_5)e^-\,.$$

Find G_V and G_A.

2-6. Consider a variant of the $SU(2)\times U(1)$ model in which the electron mixes with a heavy charged lepton, E^-. The gauge group is $SU(2)\times U(1)$ with one left-handed doublet of fermions,

ψ_L, and singlets e_R^-, E_R^- and $\mathcal{E}_R^-$ ($\mathcal{E}^-$ is not a mass eigenstate). E_R^- and $\mathcal{E}_R^-$ transform just like e_R^- under $SU(2) \times U(1)$. The Yukawa couplings are

$$-\frac{m_e}{a} \cos\xi \, \overline{e_R^-} \phi^\dagger \psi_L + \frac{m_E}{a} \sin\xi \, \overline{E_R^-} \phi^\dagger \psi_L + \text{h.c.}$$

In addition, there are gauge-invariant mass terms

$$-m_e \sin\xi \, \overline{e_R^-} \mathcal{E}_L - m_E \cos\xi \, \overline{E_R^-} \mathcal{E}_L + \text{h.c.}$$

Find ψ_L and $\mathcal{E}_L$ in terms of mass-eigenstate fields, e_L and E_L. Then calculate the decay rate for the process

$$E^- \to e^- + e^- + e^+$$

from the neutral current interaction. Neglect m_e compared to m_E.

Note: Try to get the answer from the μ decay rate actually calculating anything new.

2-7. Calculate the differential cross section $d\sigma/d\theta$ for $e^+e^- \to \mu^+\mu^-$, in the center of mass system where θ is the angle between the momentum of the incoming e^- and the outgoing μ^-. Assume that the energy is large compared to m_μ (so that the lepton masses can be ignored), but not large compared to M_Z (so that only electromagnetic and weak-electromagnetic interference terms are relevant). Further assume that the incoming es are unpolarized and the polarization of the outgoing μs is unmeasured. Integrate $d\sigma/d\theta$ to the contribution of the weak electromagnetic interference to the total cross section, σ. Find the front-back asymmetry defined by

$$\frac{1}{\sigma}\left[\int_0^1 \frac{d\sigma}{d\theta}\, d(\cos\theta) - \int_{-1}^0 \frac{d\sigma}{d\theta}\, d(\cos\theta)\right]$$

2-8. Consider a model of electrons and muons based on an $SU(3)$ gauge symmetry in which the left-handed fermion fields are a pair of $SU(3)$ 3's,

$$\psi_{1L} = \begin{pmatrix} \nu_{eL} \\ e_L^- \\ \mu_L^+ \end{pmatrix} \qquad \psi_{2L} = \begin{pmatrix} \nu_{\mu L} \\ \mu_L^- \\ e_L^+ \end{pmatrix}$$

where e^+ and μ^+ are the charge conjugated fields, $\mu^+ = (\mu^-)_c$ and $e^+ = (e^+)_c$. Under an infinitesimal $SU(3)$ transformation, the transform as follows:

$$\delta\psi_{jL} = i\epsilon_a T_a \psi_{jL}$$

where T_a are the Gell-Mann matrices. Show that this $SU(3)$ contains the usual $SU(2) \times U(1)$. Calculate $\sin^2\theta$ in tree approximation. $SU(3)$ can be broken down to $SU(2) \times U(1)$ by an octet field, ϕ (a traceless Hermitian 3×3 matrix field, transforming as $\delta\phi = i\epsilon_a[T_a, \phi]$) with VEV

$$\langle\phi\rangle = \begin{pmatrix} M & 0 & 0 \\ 0 & M & 0 \\ 0 & 0 & -2M \end{pmatrix}$$

Show that the $SU(3)$ gauge bosons that have mass of order M mediate, among other things, μ^- decay into right-handed electrons. How large does M have to be so that the rate for this decay is less than 1% of the normal decay rate?

3 — Quarks and QCD

In this section, we will give a semi-historical discussion of quantum chromodynamics, and the weak interactions of the quarks. This is only the beginning of our discussion of the weak interactions of strongly interacting particles. It is one thing to write down the model, as we will do in this chapter. It is quite another to understand what it means and whether it is right — this will occupy us for much of the rest of the book — and is one of the exciting issues in particle physics today.

3.1 Color $SU(3)$

Hadrons are built out of spin$-\frac{1}{2}$ quarks. We now have a rather convincing, though not fully quantitative, theory based on an $SU(3)$ gauge theory. the quarks and their fields (I'll call them q generically) come in three colors:

$$q = \begin{pmatrix} q_{\rm red} \\ q_{\rm green} \\ q_{\rm blue} \end{pmatrix} \tag{3.1.1}$$

In addition, the theory involves eight "gluon" fields G_a^μ, the $SU(3)$ gauge particles. The Lagrangian is

$$\mathcal{L} = -\frac{1}{4} G_a^{\mu\nu} G_{a\mu\nu} + \sum_{\rm flavors} \left(i\bar{q} \not{D} q - m_q \bar{q} q \right) \tag{3.1.2}$$

where the $G_a^{\mu\nu}$, the gluon field–strength, and D^μ, the covariant derivative, are

$$D^\mu = \partial^\mu + igT_a G_a^\mu \,, \qquad igT_a G_a^{\mu\nu} = [D^\mu,\, D^\nu] \,. \tag{3.1.3}$$

The T_a are the eight 3×3 traceless Hermitian matrices, conventionally normalized so that

$$\operatorname{tr}(T_a T_b) = \frac{1}{2}\delta_{ab} \,. \tag{3.1.4}$$

The T_a's are "color" charges just like Q in QED. The net effect of the exchange of all eight color gluons is to bind the quarks and antiquarks into color–neutral combinations, much as the photon exchange binds charged particles into electrically neutral atoms. The color–neutral combinations are

$$\bar{q}q \tag{3.1.5}$$

and

$$\epsilon_{jkl} q_j q_k q_l \tag{3.1.6}$$

where j, k, and l are color indices and repeated indices are summed. (3.1.5) describes mesons and those in (3.1.6) are baryons (of course, there are antibaryons as well). All the hadrons we have seen correspond to one of these combinations, where each q is one of the types or flavors of quarks: u (for "up"), d (for "down"), s (for "strange"), c (for "charm"), and b (for "bottom" or "beauty").

A sixth quark flavor t (for "top" or "truth") almost certainly exists and may have been seen.[1] The flavors u, c, and t have $Q = \frac{2}{3}$. Flavors d, s, b have $Q = -\frac{1}{3}$. In Table 3.1, I give the quark composition of some representative hadrons.[2]

Particle	J^p	Mass(MeV)	Quark Composition
π^+ (π^-)	0^-	139.57	$u\bar{d}$ $(d\bar{u})$
π^0	0^-	134.98	$(u\bar{u} - d\bar{d})/\sqrt{2}$
K^+ (K^-)	0^-	493.68	$u\bar{s}$ $(s\bar{u})$
K^0 $(\overline{K}^0)$	0^-	497.67	$d\bar{s}$ $(s\bar{d})$
η	0^-	547.45	$\sim (u\bar{u} + d\bar{d} - 2s\bar{s})/\sqrt{6}$
η'	0^-	957.77	$\sim (u\bar{u} + d\bar{d} - 2s\bar{s})\sqrt{3}$
D^+ (D^-)	0^-	1869.3	$c\bar{d}$ $(d\bar{c})$
D^0 $(\overline{D}^0)$	0^-	1864.5	$c\bar{u}$ $(u\bar{c})$
D_s^+ (D_s^-)	0^-	1968.5	$c\bar{s}$ $(s\bar{c})$
η_c	0^-	2979.8	$\sim c\bar{c}$
B^+ (B^-)	0^{-*}	5279	$u\bar{b}$ $(b\bar{u})$
B^0 $(\overline{B}^0)$	0^{-*}	5279	$d\bar{b}$ $(b\bar{d})$
B_s^0 (B_s^o)	0^{-*}	5369	$s\bar{b}$ $(b\bar{s})$
p	$\frac{1}{2}^+$	938.77	ddu
n	$\frac{1}{2}^+$	939.57	ddu
Λ	$\frac{1}{2}^+$	1115.68	uds
Σ^+	$\frac{1}{2}^+$	1189.37	uus
Σ^0	$\frac{1}{2}^{+*}$	1192.55	uds
Σ^-	$\frac{1}{2}^+$	1197.44	dds
Ξ^0	$\frac{1}{2}^{+*}$	1314.90	ssu
Ξ^-	$\frac{1}{2}^{+*}$	1321.32	ssd
Ω^-	$\frac{3}{2}^{+*}$	1672.45	sss
Λ_c^+	$\frac{1}{2}^{+*}$	2284.9	udc
Λ_b^0	$\frac{1}{2}^{+*}$	5641	udb

Table 3.1: Quark content of hadrons.

In perturbation theory, the QCD Lagrangian describes a world of quarks and massless gluons coupled to the eight color changes. This world doesn't look much like our own. The quarks carry fractional charges. The gluons are apparently massless. And we don't see any of these things in the real world. Nevertheless, many people believe QCD is the best candidate yet proposed for a theory of the strong interactions. Why doesn't the world look like a world of quarks and gluons?

The QCD Lagrangian with no quarks or with massless quarks has no dimensional parameter, only the dimensionless coupling constant g. Classically, therefore, the theory is scale–invariant. But to define the quantum theory, it is necessary to break the scale invariance and define or renormalize the coupling g and the fields at reference momenta of the order of a "renormalization mass: μ. Because the theory is renormalizable, the renormalization mass is completely arbitrary. A change in μ can be compensated by redefining g and rescaling the fields. In this way, we can describe the

[1]The t, however, is so heavy that it decays in to W^+b so quickly that it hardly makes sense to talk about it being confined in hadrons.

[2]* These are quark model predictions. J_p is not yet measured.

same physics with the theory defined at any μ we choose. Conversely, if the physics is fixed, the coupling constant g becomes a function of μ, $g(\mu)$. Furthermore, if $g(\mu)$ is small for some particular μ, we can find the dependence of g on μ in some region using perturbation theory by expanding in powers of the small coupling constant. When we do that in the theory with no quarks, we find (to lowest order)

$$\alpha_s(\mu) \equiv \frac{g(\mu)^2}{4\pi} = \frac{2\pi}{11} \frac{1}{\ln(\mu/\Lambda)} \tag{3.1.7}$$

Notice that dimensional transmutation has taken place. The quantum theory is characterized not by a dimensionless parameter, but by the dimensional parameter Λ. Dimensional transmutation does not depend on the perturbation formula (3.1.7). In general, $\alpha_s(\mu) = f(\mu/\Lambda)$ were $f(x)$ is some fixed function.

(3.1.7) also exhibits the property of asymptotic freedom, which means formally that $\alpha_s(\mu)$ decreases as the renormalization point μ increases. Roughly speaking, $\alpha_s(\mu)$ is the strength of the QCD interactions between color charges separated by a distance of the order of $1/\mu$. Since we can turn asymptotic freedom on its head and see that $\alpha_s(\mu)$ increases as μ decreases, this suggests that something interesting happens for μ small, which is at large separations of color charges. The best guess we have at the moment for what happens to the theory at long distances is that it confines color.

What confinement means is that we can never see a completely isolated colored quark. If we could somehow pull a single quark away from an antiquark without creating quark–antiquark pairs, then when we got to distances large compared to the fundamental length in the theory that is $1/\Lambda$, the theory is that the quark would be subject to a constant restoring force of the order of Λ^2. In other words, the potential energy of the quark–antiquark system would rise linearly with the separation r, like $\Lambda^2 r$. The constant force confines quarks into systems of size $\sim 1/\Lambda$ (or smaller). These are the hadrons.

It is sometimes said that it is quark confinement that makes QCD complicated — that makes it difficult to see quarks and gluons directly. Rubbish! Actually, what makes QCD complicated is the fact that the lightest quarks are, in a sense which we will discuss more precisely later, much lighter than the proton that is built out of them.

It is true that confinement means that it is impossible to isolate a quark. But then, we cannot isolate an electron either. After all, every electron comes complete with its long–range electromagnetic field. Any attempt to isolate an electron without its electromagnetic field is doomed because the electromagnetic force has infinite range. If our concept of "observation of a particle" is not flexible enough to apply to objects that cannot be entirely isolated, then it is not very useful.

The analogy between confinement and electromagnetism is imperfect, to be sure. The electromagnetic force, while it is long–range, at least falls off with distance, whereas we believe that the chromodynamic force between color charges goes to a constant at large distances. It is even more impossible to isolate quarks than it is to isolate charged particles. To make sure that we are not fooling ourselves with this analogy and convince ourselves that confinement is not the whole answer to the question of why we don't see quarks, let's do a little thought experiment and build a toy world. For convenience, we'll try to make this toy world look as much as possible like our own, but we'll leave open the possibility of adjusting the strength of the QCD force.

3.2 A Toy Model

For simplicity, we'll deal with a world with only two types of quarks, a μ quark with electric charge $+\frac{2}{3}$ and a d quark with charge $-\frac{1}{3}$. We'll take them approximately degenerate and give both a mass $\sim m_q$. That's a reasonable approximation to our world because it wouldn't make much difference to our everyday lives if strange quarks and charmed quarks, and heavier quarks didn't exist. We will also include in our toy world an electron, with nonQCD interactions, just ordinary electromagnetism, and a mass m_e.

The QCD interactions in this toy world, as in our own, are such that the quarks tend to bind together into color singlet states — states with no color charges. This is nothing complicated; it happens for the same reasons that electromagnetic charges tend to bind together to form neutral systems. In particular, we can build a proton state ($P = uud$) as a color singlet combination of two u quarks and a d quark and a neutron state ($N = ddu$) as a color singlet combination of two d quarks and a u quark, and these will be the lightest baryons.

In the microscopic toy world, that is in the Lagrangian, there are four parameters: m_e, the mass of the quarks; Λ, including dimensional transmutation of the QCD coupling; and α, the electromagnetic coupling. To make this look as much like our world as possible, let's take m_e and α to have the same values that they have in our world. Furthermore, let's adjust the scale of m_q and Λ so that the proton mass m_p in the toy world is the same at it is in our world. After all this, we still have one parameter left, the ratio of the dimensional parameter Λ, which characterizes the coupling constant α_s, to the quark mass m_q.

So far, we haven't done anything to make the toy world any different from our own. But that comes next. We will adjust the ratio Λ/m_q so that the QCD coupling at the proton mass is electromagnetic in size:

$$\alpha_s(m_p) = \alpha \tag{3.2.1}$$

This makes quantum chromodynamics as easy as quantum electrodynamics. Perturbation theory in QCD should be just as good and useful and accurate as perturbation theory in QED,

Now we can easily find the quark mass. Since the chromodynamic forces are weak (like electromagnetic forces), the proton is a nonrelativistic bound state. The proton mass is just three times the quark mass plus binding corrections of the order of αm_q. Thus, the quark mass is

$$m_q \simeq m_p/3 \simeq 310 \text{ MeV} \tag{3.2.2}$$

The proton radius in the toy world is a Bohr radius, $(\alpha m_q)^{-1} \simeq 10^{-11}$ cm. Here is a major difference between our world and the toy world. The toy nucleus is one hundred times bigger than the nuclei in our world. Nevertheless, it is still very small compared to the atom as a whole, so the chemistry of the toy world is probably just like what we are used to. Exactly what happens to nuclear physics in the toy world is not at all clear. Indeed, there are many peculiar things about nuclear physics in the toy world. For example, the baryon resonances Δ^{++} and Δ^- are stable if we neglect the weak interactions.

At any rate, it is at least possible to imagine that people (or some higher form of life) may exist in the toy world, and we can then ask — What will they think about quark confinement? The most likely answer is that they will not think about it very much at all. They will be perfectly satisfied with their perturbative theory of quantum chromodynamics. Perhaps some very smart creature will discover asymptotic freedom and speculate about what happens to the theory at very large distances.

To see what kind of distances we are talking about, let's calculate $1/\Lambda$. That is roughly the distance to which we must separate a quark from an antiquark before the confining force becomes important.

$$\frac{1}{\Lambda} = \exp\left(\frac{2\pi}{11\alpha}\right)\frac{1}{m_p} \simeq 10^{20}\text{cm} \tag{3.2.3}$$

In (3.2.3), I have written (3.1.7) for $\alpha_s(m_p)$ and inserted (3.2.1). Because of the exponential dependence on $1/\alpha$, Λ is tiny, and $1/\Lambda$ is a truly enormous distance. Obviously, there is no hope of seeing any confinement effects in terrestrial laboratories.

Elementary particle physics looks deceptively familiar in the toy world. For example, look at the mass spectrum of the hadrons, by which I mean color–singlet bound states of quarks and gluons. All of the familiar quark model states are there, not including strange and charmed particles because we have not included s and c quarks. In fact, the qualitative correspondence between the lightest three quark states in this simple model with the lightest baryon states in our world is quite remarkable. We do not understand quite why it should work as well as it does, and this is one of the things we will discuss in later chapters. But in the toy world, the quark model is not just a qualitative guide to the physics, it is the whole story. The Δ^{++} and Δ^- are stable, as mentioned before, because they are split from P and N only by a mass of the order αm_q and cannot decay by pion emission (the $\Delta^+(\Delta^0)$ can decay electromagnetically into $P+\gamma(N+\gamma)$ or chromodynamically into $P+$ Gluons ($N+$ Gluons)). There is a whole sequence of "'atomic" baryon resonances above the P and N by mass splittings of the order of αm_q.

There are $\overline{q}q$ bound states corresponding to pseudoscalar and vector mesons. Indeed, if the u and d quarks are exactly degenerate, it makes sense to talk about a π and an "η" (an isospin–singlet bound state of $\overline{u}u + \overline{d}d$). But in practice, if the u and d quarks differ at all in mass, isospin is irrelevant. For example, if $m_e - m_\mu$ is a few MeV (as in our world), the mesons are predominantly $\overline{u}u$ and $\overline{d}d$ bound states instead of neutral states of definite isospin. There is significant mixing between the two states because of perturbative QCD interactions (both can annihilate into gluons) that are comparable to QED interactions (annihilation into photons). At any rate, these meson states and their excitations all have a mass $\sim 2m_q \simeq 2m_p/3$. There is no light pion.

But there are light particles in the toy world. There are glueball states. The glueball states would exist in the world independent of whether there are quarks, so their mass must be of order of Λ, the only parameter in the quarkless theory. And Λ is a tiny mass, around 10^{-30} GeV. Needless to say, no one is going to measure the glueball mass in the toy world.

It is easy, in this toy world, to observe quarks. Just make them and look at them. For example, if we collide an electron and a positron, some of the time they will annihilate and produce a quark and an antiquark. The quarks will separate, and we can see them in our bubble chamber, fractional charge and everything.

We can almost see gluons the same way. Some of the time in e^+e^- annihilation, the $\overline{q}q$ pair will bremsstrahlung off a hard gluon. We can detect this thing by its collisions with nuclei. It may be that the hard gluon would pick up a very soft gluon and bind it into a color–singlet state, but we won't be able to tell the difference between a massless gluon and a glueball with a mass of 10^{-30} GeV.

I hope I have convinced you with this little fantasy that it is possible to see confined quarks. Of course, what makes the quarks visible is the same thing that makes the toy world so different from our own. The ratio of Λ to the quark mass m_q is very small; on the natural scale of the QCD interactions, the quarks are very heavy.

Now let's see what happens as we increase the ratio Λ/m_q to make the world look more like our own. The first thing that happens as we increase Λ in that confinement effects become measurable in the laboratory, but the microscopic properties are essentially the same as in the toy world. A turning point of sorts occurs for Λ of the order of 1 MeV, so confinement effects are important at atomic distances. At nuclear distances, confinement is still irrelevant, so particle physics is still well described by QCD perturbation theory.

A more important change occurs for $\Lambda \simeq \alpha_s m_q$. At this point confinement effects become important inside hadrons. This is the smallest value of Λ/m_q for which we could reasonably say that the quarks are unobservable, because here we cannot pull quarks far enough outside hadrons to see them in a nonhadronic environment. Still, we would be on shaky ground because the structure of elementary particles is still beautifully described by the nonrelativistic quark model. We just have to include a confining term in our quark–quark force law. By this time, it makes sense to talk about isospin as an appropriate symmetry, because Λ is much larger than the few–MeV mass difference between the d and u quarks. The η is now measurably heavier than the π.

Perhaps the next most important landmark is the point at which the pion becomes the lightest particle in the theory. The hadrons are still qualitatively described by a nonrelativistic bound–state picture, but now the spin–spin interactions that split the pion from the p are becoming important and bring down the pion mass. Eventually, the pion becomes lighter than the lowest–mass glueball state. The general downward trend of the pion mass as m_q/Λ decreases is correctly predicted by the quark model, but for $\Lambda > m_q$, the pion is really a rather highly relativistic bound state. In fact, we believe that as m_q goes to zero, the pion mass goes to zero, and the pion becomes the Goldstone boson associated with the spontaneous breakdown of chiral $SU(2) \times SU(2)$ symmetry. This is related to the fact that for $\Lambda > m_q$, the proton mass is considerably greater than the sum of the masses of its constituent quarks. The excess proton mass is the effect of confinement. Essentially, it is the zero–point energy of quarks confined in a region of size $1/\Lambda$. When $m_q = 0$, the proton mass is entirely due to this zero–point energy.

A bit further down in m_q/Λ, the pion becomes so light that all glueball states are unstable against decay into pions. By now, the nonrelativistic picture of hadrons has broken down badly.

In our world the pion is by far the lightest particle. The quark mass is negligible compared to Λ. Our world is not so different from the world in which the quark mass is exactly zero. Certainly we are much closer to $m_q = 0$ than to the toy world with very small Λ that we considered at the beginning.

The moral is that the reason that our world doesn't look like a world of quarks and gluons is not simply that the quarks are confined, but that the confined u and d quarks are light compared to Λ.

Despite the fact that our world is very different from the toy world in which the QCD interactions are weak, the simple nonrelativistic quark model still gives a useful description of the low–lying mesons and baryons. The quark–model picture is a useful complement to the theory of spontaneously broken chiral symmetry that we will further explore in later chapters. There is a sense, apparently, in which the u and d quarks do have a mass of a few hundred MeV, $\sim m_p/3$, with the s quark about 150 MeV heavier. These are not masses that appear in the renormalized QCD Lagrangian. They are "constituent quark masses", which include the effect of confinement, chiral symmetry breaking, or whatever.

Though its phenomenological success is undeniable, the quark model is still not understood from first principles. We have no satisfactory way, for example, of understanding the relation between the picture of the pion as a bound state of quarks with constituent masses and the complementary picture of the pion as a Goldstone boson. We will have more to say about this important puzzle in Chapters 5 and 6.

3.3 Quark Doublets

The quarks were not discovered all at once. In fact, the quark picture of hadrons grew up in the same period as the $SU(2) \times U(1)$ model. In describing the weak interactions of the quarks, I will adopt a partially historical view in which we assume the $SU(2) \times U(1)$ form and imbed the quarks in it in order of increasing mass. This is not exactly what happened, but it will preserve some of the flavor of the history without incorporating all of the confusion.

The two lightest quarks are the u and d. They are the major constituents of normal nuclei. The two left-handed quarks could be put into a $SU(2)$ doublet

$$\psi_L = \begin{pmatrix} u \\ d \end{pmatrix}_L \tag{3.3.1}$$

like the left-handed leptons. The charged–current interactions caused by $W^\pm$ exchange would then include the β–decay interaction

$$d \to u + e^- + \overline{\nu}_e \tag{3.3.2}$$

This same interaction causes π^+ decay, because the $u\overline{d}$ can annihilate into a W^+ and decay into $e^+\nu_e$ or $\mu^+\nu_\mu$. The electron mode is actually strongly suppressed in π decay because the electron mass is small compared to the pion mass. The decay is forbidden by angular–momentum conservation in the limit of zero lepton masses. Because of the $1-\lambda_5$ factor in the current, the e^+ wants to come out right-handed and the ν_e left-handed, which is impossible. The decay takes place because the e^+ mass term allows transitions from right-handed to left-handed e^+, but the amplitude is then suppressed by a factor of m_e.

(3.3.1) is inadequate to describe the interactions of the strange quark. But with two quarks with charge $-\frac{1}{3}$, it becomes possible to generalize (3.3.1) so that the field that appears in the doublet with the u quark is a linear combination of d and s.

$$\psi_L = \begin{pmatrix} u \\ \cos\theta_c d + \sin\theta_c s \end{pmatrix}_L \tag{3.3.3}$$

where θ_c is a phenomenologically important parameter called the Cabibbo angle. For a doublet of the form (3.3.3), the strangeness–conserving processes have strength $\cos\theta_c$ while the analogous processes, in which the d quark is replaced by an s quark so that they are strangeness–changing, have strength $\sin\theta_c$. Experimentally $\sin\theta_c \sim \frac{1}{5}$, so that strangeness–changing processes are inherently some 20 times weaker ($\sim 1/\sin^2\theta_c$) than strangeness–conserving processes.

Another difference in s quark decay is that the process can produce not only leptons but also light quarks. With these tools, we can understand most of the entries in the particle data book describing K decays, at least qualitatively.

Things that we cannot understand at the moment include the preponderance of hadronic decays in various systems over the semileptonic modes that we might expect to have a similar strength. For example, the rate for $\Lambda \to p\pi$ or $n\pi^0$ is 10^3 times larger than for $\Lambda \to pe^-\overline{\nu}$. Apparently something is enhancing the hadronic decays, probably some complicated strong interaction effect. The K_s^0 and K_L^0 are clearly peculiar. As we will see, these are different coherent combinations of K^0 and $\overline{K}^0$.

3.4 GIM and Charm

So far, we have discussed only $W^\pm$ exchange, but if (3.3.3) were the whole story, there would be decay processes mediated by Z^0 exchange as well. For example, the Z^0 coupling (from (2.2.21))

$$\sin\theta_c \cos\theta_c Z^0 \overline{d}_L s_L \tag{3.4.1}$$

can give rise to the decay

$$K_L^0 \to \mu^+\mu^- \tag{3.4.2}$$

at a rate comparable to the decay rate for

$$K^+ \to \mu^+ \nu_\mu \tag{3.4.3}$$

In fact, (3.4.2) is suppressed by a factor of 10^{-8}!

This is one of the symptoms of a very general problem — the absence of flavor–changing neutral current effects. The problem is particularly acute in the $SU(2)\times U(1)$ theory where there are neutral current effects in tree approximation, but it is present in almost any theory of the weak interactions.

The problem prompted Glashow, Iliopoulos, and Maiani (GIM) to propose a radical solution, the existence of a new quark, the charmed quark in an $SU(2)$ doublet with the orthogonal combination of s and d, so the doublet is

$$\psi'_L = \begin{pmatrix} c \\ \cos\theta_c\, s - \sin\theta_c\, d \end{pmatrix}_L \tag{3.4.4}$$

Now the Z^0 coupling to this doublet precisely cancels the strangeness–changing coupling from the other doublet. This is easy to see directly. But we can also see it indirectly by rotating ψ_L and ψ'_L into another so that the Cabibbo mixing is entirely on the c and u fields.

This eliminates (3.4.2) in tree approximation. As we will see in Chapter 10, the problem will reappear in one loop if the charmed quark is too heavy. The charmed quark had appear with a mass of less than a few GeV. And it did! The D^+, D^0 and Λ_c^+ have all been observed with just the right properties. The charmed quark should decay primarily into s because of (3.4.4), and it does.

This was one of the great triumphs of theoretical physics. It made a tidy little world with a correspondence between quarks and leptons — one light family with u, d, ν_e, e^- and one heavy family with c, c, ν_μ, μ^-. Unfortunately, nature didn't stop there. The τ^- lepton ruined the correspondence, and it didn't come as much of as a surprise when a b quark was later discovered. We now know that there is a t quark. God, I suppose, knows whether there will be any more. There is some evidence (which we will discuss later) from the total width of the Z that we really have now seen all the matter particles. We will assume this to be so, and in the next section we will analyze the $SU(2) \times U(1)$ theory for six quarks.

3.5 The Standard Six–Quark Model

We are now ready to embed the quarks in $SU(2) \times U(1)$. We will assume that there are six quarks and no more. To get charged–current weak interactions involving only the left-handed quarks, we must assume that the left-handed quarks are in doublets while the right-handed quarks are in singlets, just as for the leptons.

We will write down the most general theory coupling the quark and lepton fields to the Higgs doublet. Let's start by inventing a better notation. Call the lepton doublets

$$L_{jL} = \begin{pmatrix} \nu_j \\ l_j^- \end{pmatrix}_L \tag{3.5.1}$$

where j runs from 1 to 3 and labels the fields in a completely arbitrary way. The right-handed singlets are

$$l_{jR}^- \tag{3.5.2}$$

They satisfy

$$\begin{aligned} \vec{T} L_{jL} &= \frac{\vec{\tau}}{2} L_{jL}, & S L_{jL} &= -\frac{1}{2} L_{jL} \\ \vec{T} l_{jR}^- &= 0, & S l_{jR}^- &= -l_{jR}^- \end{aligned} \tag{3.5.3}$$

The most general $SU(2) \times U(1)$ invariant Yukawa coupling of these fields to the Higgs field ϕ is

$$-f_{jk}\overline{l_{jR}^-}\phi^\dagger L_{kL} + \text{h.c.} \tag{3.5.4}$$

where f_{jk} is a completely arbitrary complex matrix. When ϕ gets its VEV, (2.7.13), this becomes the charge–lepton mass term

$$-f_{jk}\frac{v}{\sqrt{2}}\overline{l_{jR}^-}l_{kL}^- + \text{h.c.} \tag{3.5.5}$$

This doesn't look like much, but we can use the fact that the kinetic–energy term has separate $SU(3) \times U(1)$ global symmetries for the l_{jR}^- and the L_{jL}^-. As we discussed in (1.2.45)–(1.2.47), we can redefine the l_R^- and L_L^- so that the mass matrix is real, diagonal, and positive. Thus, quite generally, the L_L^- and l_R^- can be taken to be pure mass eigenstate fields. We have not mentioned neutrino masses. Later on we will discuss the possibility that neutrinos have a very small mass but for the moment we will assume they are massless. Then the neutrino fields are all identical, and the ν_e, for example, can simply be taken to be the field that is in a doublet with the e_L^-. Thus, the theory of leptons described in Chapter 2 is just the most general broken $SU(2) \times U(1)$ of the L's and ϕ. Let us then see what is the most general theory of quarks.

Call the left-handed quark doublets

$$\psi_{jL} = \begin{pmatrix} U_j \\ D_j \end{pmatrix}_L \tag{3.5.6}$$

where U_j are the charge $\frac{2}{3}$ fields, and the D_j are the charge $-\frac{1}{3}$ fields. The right-handed fields are singlets

$$U_{jR} \text{ and } D_{jR} \tag{3.5.7}$$

The $SU(2) \times U(1)$ charges are

$$\begin{aligned} \vec{T}\psi_{jL} &= \frac{\vec{\tau}}{2}\psi_{jL}, \qquad S\psi_{jL} = \frac{1}{6}\psi_{jL} \\ \vec{T}U_{jR} &= \vec{T}D_{jR} = 0 \\ SU_{jR} &= \frac{2}{3}U_{jR}, \qquad SD_{jR} = -\frac{1}{3}D_{jR} \end{aligned} \tag{3.5.8}$$

The Yukawa couplings responsible for the charge $-\frac{1}{3}$ mass is

$$-g_{jk}\overline{D}_{jR}\phi^\dagger\psi_{kL} + \text{h.c.} \tag{3.5.9}$$

When ϕ develops a VEV, this produces the mass term

$$-M_{jk}^D\overline{D}_{jR}D_{kL} + \text{h.c.} \tag{3.5.10}$$

where

$$M_{jk}^D = g_{jk}v/\sqrt{2} \tag{3.5.11}$$

It is easy to check that (3.5.9) is invariant under $SU(2) \times U(1)$ by comparing (2.7.2) and (3.5.8).

To construct a Yukawa coupling involving the U_{jR} field, we need to construct a modified Higgs field with the opposite S value. We define

$$\tilde{\phi} = i\tau_2\phi^* \tag{3.5.12}$$

It is easy to see that (2.7.4) implies

$$\delta\tilde{\phi} = i\epsilon_a \frac{\tau_a}{2}\tilde{\phi} - \frac{i}{2}\epsilon\tilde{\phi} \tag{3.5.13}$$

so that

$$\vec{T}\tilde{\phi} = \frac{\tau}{2}\tilde{\phi}, \quad S\tilde{\phi} = -\frac{1}{2}\tilde{\phi} \tag{3.5.14}$$

Then we can write the invariant Yukawa coupling

$$-h_{jk}\overline{U}_{jR}\tilde{\phi}^{\dagger}\psi_{kL} + \text{h.c.} \tag{3.5.15}$$

Of course, the $\tilde{\phi}$ field is constructed so that the neutral element is on top of the doublet

$$\tilde{\phi} = \begin{pmatrix} \phi^{0*} \\ \phi^{-} \end{pmatrix}, \quad \phi^{-} = -\phi^{+*} \tag{3.5.16}$$

Its VEV then produces the U term

$$-M^U_{jk}\overline{U}_{jR}U_{kL} + \text{h.c.} \tag{3.5.17}$$

$$M^U_{jk} = h_{jk}v/\sqrt{2} \tag{3.5.18}$$

Now we can use the symmetries of the kinetic–energy term to simplify the mass terms. Each of the three types of quark fields, U_R, D_R, and ψ_L can be redefined by an $SU(2)\times U(1)$ unitary transformation. For example, we can redefine the fields so that the M^U mass term is diagonal

$$-M^U_{jk}\overline{U}_{jR}U_{kL} \tag{3.5.19}$$

$$M^U = \begin{pmatrix} m_u & 0 & 0 \\ 0 & m_c & 0 \\ 0 & 0 & m_t \end{pmatrix} \tag{3.5.20}$$

But now we cannot diagonalize M^D because we have already fixed the ψ_L field. We still have freedom to redefine the D_R fields so that we can write the D mass term as

$$-M^D_{jl}V^{\dagger}_{lk}\overline{D}_{jR}D_{kL} + \text{h.c.} \tag{3.5.21}$$

$$M^D = \begin{pmatrix} m_d & 0 & 0 \\ 0 & m_s & 0 \\ 0 & 0 & m_b \end{pmatrix} \tag{3.5.22}$$

and

$$V^{\dagger}\text{is a unitary matrix} \tag{3.5.23}$$

The D_{jR} are now simply mass eigenstate fields, but the D_{kL} are related to mass eigenstates by the unitary matrix $V^{\dagger}$.

$$\begin{pmatrix} d \\ s \\ b \end{pmatrix}_L = U^{\dagger}D_L \tag{3.5.24}$$

Thus

$$D_L = V\begin{pmatrix} d \\ s \\ b \end{pmatrix}_L \tag{3.5.25}$$

In terms of the mass eigenstate fields, the quark charged current is then

$$\begin{pmatrix} \overline{u} & \overline{c} & \overline{t} \end{pmatrix} \gamma^\mu (1+\gamma_5) V \begin{pmatrix} d \\ s \\ b \end{pmatrix} \tag{3.5.26}$$

The unitary matrix V is all the information that is left of the general mass matrices we started with, except for the quark masses themselves.

A general unitary matrix V has nine parameters. However, we are still not finished redefining fields. We can still redefine the phase of any mass eigenstate field. This means that we can multiply V on either side by a diagonal unitary matrix. Thus, we can eliminate some phases in V. Only five of the six phases in the two diagonal matrices are relevant, so we are left with four real parameters in V. To see how this works explicitly, let us repeat the analysis first performed by Kobayashi and Maskawa.

We can write a general unitary 3×3 matrix in a kind of complex Euler angle parametrization as follows:

$$V = V_2 V_1 V_3 \tag{3.5.27}$$

where matrices have the form

$$V_1 = \begin{pmatrix} X & X & 0 \\ X & X & 0 \\ 0 & 0 & X \end{pmatrix} \tag{3.5.28}$$

$$V_{2,3} = \begin{pmatrix} X & 0 & 0 \\ 0 & X & X \\ 0 & X & X \end{pmatrix} \tag{3.5.29}$$

We can show that by making diagonal phase redefinitions, we can make V_2 and V_3 real and write V_1 in a very simple form. First, let us show that any 2×2 unitary matrix can be made completely real. A unitary 2×2 matrix must have the form

$$w = \begin{pmatrix} \epsilon c & \epsilon \eta s \\ -\rho s & \rho \eta c \end{pmatrix} \tag{3.5.30}$$

where $|\epsilon| = |\eta| = |\rho| = 1$, $c = \cos\theta$, and $s = \sin\theta$. Then

$$\begin{pmatrix} \epsilon^* & 0 \\ 0 & \rho^* \end{pmatrix} w \begin{pmatrix} 1 & 0 \\ 0 & \eta^* \end{pmatrix} = \begin{pmatrix} c & s \\ -s & c \end{pmatrix} \tag{3.5.31}$$

Thus, we can redefine

$$V \to D V D' = D V_2 D'' \, D''^* V_1 D'''^* D''' V_s D' = V_2' V_1' V_3' \tag{3.5.32}$$

where D, $D' \ldots$ are diagonal unitary matrices, and

$$V_2' = \begin{pmatrix} 1 & 0 & 0 \\ 0 & c_2 & -s_2 \\ 0 & s_2 & c_2 \end{pmatrix} = \begin{pmatrix} 1 & 0 & 0 \\ 0 & c_s & s_3 \\ 0 & -s_3 & c_3 \end{pmatrix} \tag{3.5.33}$$

with $c_j = \cos\theta_j$, $s_j = \sin\theta_j$. V_1' is still an arbitrary matrix with the block diagonal form (3.5.28). We can still further redefine V by multiplying on either side by matrices of the form

$$\begin{pmatrix} \alpha & 0 & 0 \\ 0 & \beta & 0 \\ 0 & 0 & \beta \end{pmatrix} \qquad |\alpha| = |\beta| = 1 \tag{3.5.34}$$

which commute with V_2' and V_3'. It is easy to see that using (3.5.34), we can redefine V_1' to have the form

$$V_1' = \begin{pmatrix} c_1 & s_1 & 0 \\ -s_1 & c_1 & 0 \\ 0 & 0 & e^{i\delta} \end{pmatrix} \tag{3.5.35}$$

(3.5.32), (3.5.33), and (3.5.35) define the Kobayashi-Maskawa (KM) matrix. The interesting feature is the appearance of a single nontrivial phase δ. As we will see, if $\delta \neq 0$, the theory violates CP.

In one sense, the six–quark model represents a retreat from the level of understanding of the nature of the weak current in the four–quark model. In a four–quark model, there is automatically a relation among the strengths of strangeness–conserving quark interactions, strangeness–violating interactions, and the μ decay interaction, just because $\sin^2\theta_c + \cos^2\theta_c = 1$. This relation is called Cabibbo universality, and it is observed to be rather well satisfied. In the six–quark language, this means that s_3 is observed to be small. We don't know why it should be so small. But then, we know almost nothing about why the quark masses and angles are what they are.

Note that the six–quark KM model, like the four–quark GIM model that it generalizes, has no flavor–changing neutral currents in tree approximation. This is simply because T_3 and S commute with any matrix in flavor space. Thus, in any basis, the Z^0 couples to the quark neutral current (with $x = \sin^2\theta$)

$$\begin{aligned} \frac{e}{\sin\theta\cos\theta}\sum_j \Big\{ \Big(\frac{1}{2} - \frac{2}{3}x\Big)\overline{U}_{jL}\gamma^\mu U_{jL} + \Big(-\frac{1}{2} + \frac{1}{3}x\Big)\overline{D}_{jL}\gamma^\mu D_{jL} \\ -\frac{2}{3}x\overline{U}_{jR}\gamma^\mu U_{jR} + \frac{1}{3}x\overline{D}_{jR}\gamma^\mu D_{jR}\Big\}. \end{aligned} \tag{3.5.36}$$

It is only the mismatch between the unitary matrices required to diagonalize M^U and M^D that produces flavor–changing effects. And these effects show up only in the charged–current weak interactions that involve U and D simultaneously.

In general, the constraint that neutral currents will not change flavor in tree approximation can be stated precisely as follows: Every set of fields whose members mix among themselves must have the same values of T_3 and S.

There is no universally used convention for the angles in the KM matrix. Nor does everyone agree about where to put the nontrivial phase. However, everyone agrees about the form of the current in (3.5.26). The matrix element V_{jk} is the coefficient of the term in the current with charged 2/3 quark field $\bar{j}$ and charged $-1/3$ quark field k. Thus in the convention and obvious basis of (3.5.26), V has the form

$$V = \begin{pmatrix} V_{ud} & V_{us} & V_{ub} \\ V_{cd} & V_{cs} & V_{cb} \\ V_{td} & V_{ts} & V_{tb} \end{pmatrix} \tag{3.5.37}$$

and at least the meaning of the absolute value of the matrix elements is unambiguous.

3.6 *CP* Violation

The unremovable phase in the KM matrix is a signal of CP violation. A CP transformation takes a simple form in a special representation of the γ matrices called a Majorana representation. A Majorana representation is one in which the γ matrices are imaginary. For example, we can take

$$\begin{aligned} \gamma^0 = \sigma_2, \quad \gamma^1 = i\sigma_s\tau_1 \\ \gamma^2 = i\sigma_1, \quad \gamma^3 = i\sigma_s\tau_3 \end{aligned} \tag{3.6.1}$$

In this representation, charge conjugation is trivial, just complex conjugation.

Then the CP transformation is

$$\begin{aligned} \psi(\vec{x}, t) &\rightarrow \gamma^0 \psi^*(-\vec{x}, t) = -\overline{\psi}^T(-\vec{x}, t) \\ \overline{\psi}(\vec{x}, t) &\rightarrow \psi^T(-\vec{x}, t) \end{aligned} \tag{3.6.2}$$

Thus, it just interchanges ψ and $\overline{\psi}$ with a minus sign to make up for Fermi statistics.

A kinetic–energy term, ever for a parity–violating state like a neutrino is invariant under (3.6.2). This is why CP is interesting.

$$i\overline{\psi}\partial\!\!\!/ P_+ \psi \tag{3.6.3}$$

is not invariant under C or P separately. But under (3.6.2), because $\vec{\gamma} P_+$ is symmetric and $\gamma^0 P_+$ is antisymmetric in a Majorana basis, (3.6.3) transforms into itself plus a total divergence.

A mass term or gauge coupling can be invariant under (3.6.3) if the masses and couplings are real. In particular, consider the coupling of the W^+ to quarks. It has the form

$$g\overline{U} W\!\!\!\!/^{\,+} P_+ D + g^* \overline{D} W\!\!\!\!/^{\,-} P_+ U \tag{3.6.4}$$

The CP operation interchanges the two terms (with appropriate interchange of the W^+ fields), except that g and g^* are not interchanged. Thus, CP is a symmetry only if there is some basis in which all the couplings (and masses) are real. If there were only the four quarks u, d, c and s, the $SU(2) \times U(1)$ interactions could not violate CP. As we have seen, we *can* find a basis in which masses and couplings are real. The important result of Kobayashi and Maskawa was that with six quarks or more, the $SU(2) \times U(1)$ interactions can violate CP. It is, of course, an important experimental question whether the KM phase δ is in fact the source of observed CP violation. We will discuss this question in detail when we study K^0 physics.

Problems

3-1. Suppose the t quark didn't exist. How could you build a five-quark $SU(2) \times U(1)$ theory? What phenomenological difficulties would such a theory face if we did not know that the t existed?

3-2. Find the number of real angles and unremovable phases in the 4×4 generalization of the KM matrix.

3-3. Find a copy of the most recent Particle Physics Booklet from the particle data group, or the Review of Particle Properties. Look through the summary tables for Gauge and Higgs bosons, Mesons, and Baryons. Identify each **exclusive**[3] decay mode with a branching ratio of more than 2% that is produced by the electroweak interactions. Be prepared to talk about each such mode and draw a representative Feynman graph that contributes to the decay process.

3-4. Construct an alternative $SU(2) \times U(1)$ model of the weak interactions of the u, d, s and c quarks that involves a pair of heavy (unobserved) quarks, h and l, with charges $\frac{5}{3}$ and $-\frac{4}{3}$, respectively. One linear combination of the left-handed light-quark fields is in an $SU(2)$ doublet:

$$\psi_L = \begin{pmatrix} U_{1L} \\ D_{1L} \end{pmatrix}$$

[3] An exclusive decay is one that involves a definite final state, not something like $Z \rightarrow \Lambda X$, which refers to an inclusive decay into Λ plus anything.

The other is an $SU(2)$ quartet with the h and l:

$$X_L = \begin{pmatrix} h_L \\ U_{2L} \\ D_{2L} \\ l_L \end{pmatrix}$$

All right-handed fields are $SU(2)$ singlets. Find $T_a X$ and SX. What Higgs fields are required to give mass to these quarks. Can the $SU(2) \times U(1)$ interactions in this theory violate CP? Explain.

4 — $SU(3)$ and Light Hadron Semileptonic Decays

4.1 Weak Decays of Light Hadrons

In Chapters 2 and 3, I have written down the standard $SU(3) \times SU(2) \times U(1)$ model of the strong and electroweak (short for weak and electromagnetic) interactions of leptons and quarks. To understand the theory fully, we need to understand two more things: hadrons and loops. That is, we must understand how the weak interactions of quarks give rise to those of the hadrons. We have seen that they work qualitatively. We would like to say something quantitative. Then we would like to see the renormalizability of the theory in action by computing finite radiative corrections.

We will begin by using $SU(3)$ symmetry to analyze some weak decays of hadrons. In the process, we will determine some of the parameters in the KM matrix.

Historically, the weak decays of the hadrons are divided into three classes: leptonic decays, which have only leptons (no hadrons) in the final state, such as $\pi^+ \to \mu^+\nu_\mu$ or $K_L \to \mu + \mu^-$; semileptonic decays, which have both leptons and hadrons in the final state, such as $K^+ \to i^0\mu^+\nu_\mu$ or $n \to pe^-\overline{\nu}_e$; and nonleptonic decays, which have no leptons (only hadrons) in the final state, such as $K_s \to \pi^+\pi^-$ or $\Lambda \to p\pi^-$.

From our modern vantage point, this division misses the point. the important division is between those decays that involve leptons (which are produced by four–fermion operators involving two lepton fields and two quark fields) and the nonleptonic decays (produced by four–fermion operators involving four quark fields).

4.2 Isospin and the Determination of V_{ud}

The piece of the quark charged current involving the light u and d quarks is

$$V_{ud}(j^\mu + j_5^\mu) = V_{ud}\overline{u}\gamma^\mu(1+\gamma_5)d. \tag{4.2.1}$$

This is built out of the vector and axial isospin currents

$$j_a^\mu = \overline{\psi}\gamma^\mu \frac{\tau_a}{2}\psi \tag{4.2.2}$$

and

$$j_{5a}^u = \overline{\psi}\gamma^\mu\gamma_5\frac{\tau_a}{2}\psi, \tag{4.2.3}$$

where ψ is the isospin doublet

$$\psi = \begin{pmatrix} u \\ d \end{pmatrix}. \tag{4.2.4}$$

Because the $u-d$ quark mass difference is very small compared to Λ, the isospin charges

$$Q_a = \int j_a^0(x)\, d^3x, \tag{4.2.5}$$

almost commute with the Hamiltonian. Isospin is broken by the electromagnetic (and weak) interactions and by the $u-d$ mass difference, but it is observed to be an excellent symmetry of the strong interactions.

Of course, we believe that the u and d quark masses themselves are not much bigger than the mass differences. For the moment, we will bypass them because they are not realized in the usual way but are spontaneously broken.

Anyway, we can use symmetry arguments to determine the matrix elements of the Q_a, and this in turn should tell us about matrix elements of the currents (4.2.2).

So we would like to identify processes in which only the vector current contributes. We must look for a decay involving leptons, because the light quarks only decay into leptons and, besides, a purely hadronic decay would involve the matrix element of a product and two quark currents. The axial current contribution can be eliminated in a special case. If the decaying hadron A has spin zero and decays into another spinless hadron B with the same parity, in the same isomultiplet, the amplitude for decay is

$$V_{ud}\langle B|j^\mu + j_5^\mu|A\rangle\, l_\mu \cdot \frac{G_F}{\sqrt{2}}, \tag{4.2.6}$$

where l_μ is the lepton matrix element. But the matrix element of j_5^μ vanishes because of the parity invariance of strong interaction! There is no way to form an axial vector of only the two hadron momenta.

Thus, the matrix element of interest is that of j^μ. If the momenta of A and B are p and p', and we define the momentum transfer

$$q^\mu = p^\mu - p'^\mu, \tag{4.2.7}$$

we can write the matrix element

$$\langle B, p'|j^\mu(0)|A, p\rangle = C(q^2)(p^\mu + p'^\mu) + D(q^2)q^\mu. \tag{4.2.8}$$

This form follows simply from Lorentz invariance. The invariant functions C and D also depend, in general, on p^2 and p'^2, but since these are fixed at the particle masses, we suppress this dependence.

We want to use symmetry arguments to say something about C and D. We will, therefore, do a rather peculiar thing. We will look at the matrix element in the limit of exact isospin symmetry. The reason that this is peculiar is that in this limit A and B are degenerate, since they are in the same isospin multiplet. Thus, A doesn't decay into B in this limit! Nevertheless, we will try to say something about C and D in this limit, the extrapolate back to the real world, assuming that C and D don't change very much. Then we can use the real observed masses and calculate the decay rate.

In the symmetry limit, the current j^μ is conserved. Thus q_μ, contracted with (4.2.8) must vanish, and therefore we don't have to worry about D because

$$q^2 D(q^2) = 0 \Rightarrow D(q^2) = 0 \quad \text{for } q^2 \neq 0. \tag{4.2.9}$$

To get a handle on C, we note that the matrix element of the charge Q_a between A and B is completely determined. Since A and B are in the same multiplet, we can write

$$\begin{aligned} |A, p\rangle &= |j, m, p\rangle \\ |B, p'\rangle &= |j, m', p'\rangle, \end{aligned} \tag{4.2.10}$$

where j and $m(m')$ are the isospin and Q_3 values. In general

$$Q_a|j,\,m,\,p\rangle = |j,\,m',\,p\rangle(T_a)_{m'm}, \tag{4.2.11}$$

where T_a are the isospin matrices for isospin j. In particular, we are interested in the raising operator

$$Q_+ = Q_1 + iQ_2 = \int j^0(x)\,d^3x = \int u^\dagger(x)d(x)\,d^3x, \tag{4.2.12}$$

which satisfies

$$Q_+|j,\,m,\,p\rangle = \sqrt{(j-m)(j+m+1)}\,j,\,m+1,\,p\rangle, \tag{4.2.13}$$

so m' in (4.2.10) is $m+1$ (this is obvious — the charged–current weak interactions change the charge by one). Thus

$$\langle B,\,p'|Q_+|A,\,p\rangle = \sqrt{(j-m)(j+m+1)}\,(2\pi)^3 2p^0\delta^{(3)}(\vec{p}-\vec{p}\,'), \tag{4.2.14}$$

for states with standard invariant normalization.

Now let us calculate (4.2.14) a different way:

$$\langle B,\,p'|j^0(x)|A,\,p\rangle = e^{ix(p'-p)}\langle B,\,p'|j^0(0)|A,\,p\rangle, \tag{4.2.15}$$

from translation invariance. Thus, from (4.2.8)—(4.2.9)

$$\langle B,\,p'|Q_+|A,\,p\rangle = C(0)2p^0(2\pi)^3\delta^{(3)}(\vec{p}-\vec{p}\,'). \tag{4.2.16}$$

Note, as expected, the time dependence of Q_+ goes away in the symmetry limit because $p'^0 = p^0$ when $\vec{p}\,' = \vec{p}$. Now we can compare (4.2.14) and (4.2.16) and find

$$C(0) = \sqrt{(j-m)(j+m+1)}. \tag{4.2.17}$$

Thus, we know $C(0)$ in the symmetry limit. This should be good enough to give an excellent approximation to the decay rate. In first approximation, we will just assume that $C(q)$ is constant and equal to its symmetry value (4.2.17). The difference between $C(0)$ in the real world and in the symmetric world should not be greater than $\Delta m/\Lambda$, where Δm is the u-d mass difference, a couple of MeV, while Λ is a few hundred MeV.[1] The q^2 dependence of C should likewise be determined by Λ, while the q^2 in the decay is less than the $A-B$ mass difference squared, of the order of $(\Delta m)^2$.

Of course, if we want very accurate predictions, we can do better by modeling (or measuring) the q^2 dependence and the symmetry–breaking effects. We could also include the effect of photon exchange and other radiative corrections (now that we have a renormalizable theory). So a decay such as this can be characterized very accurately in terms of the single parameter V_{ud}.

There are many such process. In particle physics, the decay

$$\pi^+ \to \pi^0 e^+ \nu_e \tag{4.2.18}$$

has been seen. In nuclear physics, there are lots of them. For example

$$\begin{gathered} {}^{14}\mathrm{O} \to {}^{14}\mathrm{N}\,e^+\nu_e \\ {}^{34}\mathrm{Cl} \to {}^{34}\mathrm{S}\,e^+\nu_e \\ {}^{26}\mathrm{Al} \to {}^{26}\mathrm{Mg}\,e^+\nu_e. \end{gathered} \tag{4.2.19}$$

[1] In fact, the contribution from the u-d mass difference is even smaller because of the Ademollo-Gatto theorem, but we will discuss this later when we deal with strangeness changing currents.

Since we are just using symmetry arguments to get our prediction, we don't mind that the hadrons are complex nuclei instead of nice simple "elementary particles".

In fact, the most accurate data come from the nuclear decays and give, after suitable corrections, the value shown in the particle data book.

$$V_{ud} = 0.9747 \text{ to } 0.9759$$

So we know at least one of the KM parameters rather well.

4.3 f_π

Having determined V_{ud} by studying the vector part of (4.2.1), we can proceed to look at the simplest process involving the axial vector part. This is π decay

$$\pi^- \to \mu^- \overline{\nu}_\mu \text{ or } e^- \overline{\nu}_e, \tag{4.3.1}$$

with amplitude

$$\langle 0 | j_5^\mu | \pi^- \rangle \, l_\mu \frac{V_{ud} G_F}{\sqrt{2}}. \tag{4.3.2}$$

Only the axial current contributes because of parity. We are not smart enough to calculate the hadronic matrix element, but we can parametrize it:

$$\langle 0 | j_5^\mu | \pi^- \rangle = i f_\pi p^\mu, \tag{4.3.3}$$

where p^μ is the π momentum, the current is

$$j_5^\mu = \overline{u} \gamma^\mu \gamma_5 d, \tag{4.3.4}$$

and f_π is a constant with dimensions of mass. The π decay rate can then be calculated in terms of f_π so that f_π can be determined from the observed rate. It is

$$f_\pi \simeq 0.93 \, m_\pi. \tag{4.3.5}$$

Be warned that there is another perfectly reasonable definition of f_π that appears in the literature, in terms of the self-adjoint isospin currents and fields. It is

$$\langle 0 | j_{5a}^\mu | \pi_b \rangle = i \delta_{ab} F_\pi p^\mu. \tag{4.3.6}$$

I use capital F to distinguish it from (4.3.3), but this convention is not consistently used. The literature is random. The relation between these two definitions is

$$i f_\pi p^\mu = \left\langle 0 \,\middle|\, j_{51}^\mu + i j_{52}^\mu \,\middle|\, \frac{\pi_1 - i\pi_2}{\sqrt{2}} \right\rangle = i\sqrt{2} \, F_\pi p^\mu. \tag{4.3.7}$$

Thus

$$F_\pi = f_\pi / \sqrt{2} \simeq 93 \, \text{MeV}. \tag{4.3.8}$$

Notice that if we tried to integrate the current in (4.3.2) over space to get a charge, we would have the charge acting on the vacuum to produce not zero, but a zero–momentum pion state. If this situation persists in the symmetry limit $m_u = m_d = 0$, it is an indication of spontaneous symmetry breaking, since the charge doesn't annihilate the vacuum state. Another way to see this is to note that in the symmetry limit, the j_5^μ current is conserved. Thus, if we contract (4.3.3) with p_μ, we must get zero, so

$$f_\pi p^2 = 0 = f_\pi m_\pi^2. \tag{4.3.9}$$

Either f_π vanishes, or $m_\pi^2 = 0$ and the pion is the Goldstone boson associated with spontaneous breakdown of the chiral isospin symmetry. We believe that it is the latter and that our world is very close to a world with a Goldstone pion.

4.4 Strangeness Changing Currents

Now that we have determined the coefficient of the piece of the quark current involving u and d quarks, we turn our attention to the next lightest quark and look for the term

$$V_{us}\overline{u}\gamma^\mu(1+\gamma_5)s. \tag{4.4.1}$$

We can measure V_{us} (perhaps) by looking at the strangeness–changing semileptonic decays of strange particles (since the current destroys an s quark and replaces it by a u quark).

The simplest such process that we can hope to calculate from symmetry principles is the K_{l3} decays:

$$\begin{gathered} K^- \to \pi^0 e^- \overline{\nu}_e \\ K^- \to \pi^0 \mu^- \overline{\nu}_\mu \\ K_L^0 \to \pi^+ e^- \overline{\nu}_e \\ K_L^0 \to \pi^+ \mu^- \overline{\nu}_\mu \end{gathered} \tag{4.4.2}$$

We might hope to calculate these just as we calculated $\pi^- \to \pi^0 e^- \overline{\nu}_e$. The amplitude for K^- decay, for example, is

$$\langle \pi^o | j^\mu | K^- \rangle l_\mu \cdot V_{us} G_F/\sqrt{2}, \tag{4.4.3}$$

where j^μ is $j^\mu = \overline{u}\gamma^\mu s$. The charge associated with j^μ is a generator of $SU(3)$, which is a good symmetry in the limit that the u, d, and s quarks are all degenerate. As in (4.2.8)m we can write

$$\langle \pi^0 | j^\mu | K^- \rangle = f_+(q^2)(p_\pi^\mu + p_K^\mu) + f_(q^2)(p_K^\mu - p_\pi^\mu), \tag{4.4.4}$$

where $q^\mu = p_K^\mu - p_\pi^\mu$. As before, in the symmetry limit, we can argue that $f_-(q^2) = 0$, and we can calculate $f_+(0)$ from symmetry arguments. Then if we ignore the symmetry breaking and the q^2 dependence of f_+, we can calculate the decay rate and compare with experiment. This determines V_{us}. The result is $0.216 \pm .003$, as calculated by R. E. Schrock and L. L. Wang (*Phys. Rev. Lett.* **41**:1692–1695, 1978).

We may not believe this result! $SU(3)$ symmetry is not as good as $SU(2)$. Furthermore, $SU(3)$ breaking in the pseudoscalar meson sector seems to be especially severe. The K-π mass difference is more than twice as large as the π mass. However, the situation is not quite so bad as it seems. As we will see later, the large K-π mass difference is a consequence of the fact that both the π and the K are almost Goldstone bosons — it is the difference in the squared masses, $m_K^2 - m_\pi^2$ that is proportional to the $SU(3)$ breaking quark mass difference, $m_s - m_d$ (or $m_s - m_u$). Furthermore, the Ademollo-Gatto theorem[2] implies that the symmetry breaking effect on the matrix elements of the current is actually quadratic in the symmetry breaking — the linear terms cancel. The physics of the Ademollo-Gatto theorem is that any symmetry breaking that effects the vector current also effects the derivative terms in the Lagrangian. For symmetry breaking due to the quark masses, the first order effect on the derivative terms changes the wave-function renormalization of the fields in the same way it changes the currents, so the symmetry breaking to this order can be removed by redefining the fields and there is no physical effect in first order. We will be able to see this easily later, in chapter 6, when we explicitly write down the chiral Lagrangian that describes the πs and Ks.

[2] M. Ademollo and R. Gatto, Phys. Rev. Lett. **13** 264, 1964.

It worth looking at other places to get a handle on V_{us}. $SU(3)$ breaking in the baryon family, the octet with spin$-\frac{1}{2}$, seems much less violent. Furthermore, there are several semileptonic decays we can look at.

$$\begin{aligned}
n &\to pe^-\overline{\nu}_e \\
\Lambda &\to pe^-\overline{\nu}_e \quad \text{or} \quad p\mu^-\overline{\nu}_\mu \\
\Sigma^+ &\to \Lambda e^+\nu_e \\
\Sigma^- &\to ne^-\overline{\nu}_e \quad \text{or} \quad n\mu^-\overline{\nu}_\mu \quad \text{or} \quad \Lambda e^-\overline{\nu}_e \\
\Xi^- &\to \Lambda e^-\overline{\nu}_e \quad \text{or} \quad \Lambda\mu^-\overline{\nu}_\mu
\end{aligned} \tag{4.4.5}$$

This looks like a good system to study.

4.5 *PT* Invariance

We are thus led to consider the matrix element of the charged weak current between baryon states. Since both the vector and axial vector parts contribute, we must evaluate both

$$\langle B',\, p',\, s'|j^\mu|B,\, p,\, s\rangle = \overline{u}(p',\, s')\,[f_1\gamma^\mu - if_2\sigma^{\mu\nu}q_\nu + f_3q^\mu]\,u(p,\, s) \tag{4.5.1}$$

and

$$\langle B',\, p',\, s'|j_5^\mu|B,\, p,\, s\rangle = \overline{u}(p',\, s')\,[g_1\gamma^\mu\gamma_5 - ig_2\sigma^{\mu\nu}\gamma_5q_\nu + g_3\gamma_5q^\mu]\,u(p,\, s) \tag{4.5.2}$$

where f_i and g_i depend only on $q^2 = (p - p')^2$. These forms follow from Lorentz invariance and parity invariance of the strong interactions. The other tensor forms such as $p^\mu - p'^\mu$ can be reduced to these using the Dirac algebra. For example, when between $\overline{u}(p',\, s')$ and $u(p,\, s)$

$$\begin{aligned}
p^\mu + p'^\mu &= \tfrac{1}{2}\{\not p,\, \gamma^\mu\} + \tfrac{1}{2}\{\not p',\, \gamma^\mu\} \\
&= \gamma^\mu \not p + \tfrac{1}{2}[\not p,\, \gamma^\mu] + \not p'\gamma^\mu - \tfrac{1}{2}[\not p',\, \gamma^\mu] \\
&= (m + m')\gamma^\mu + i\sigma^{\mu\nu}q_\nu
\end{aligned} \tag{4.5.3}$$

Note that the spin polarization vectors s^μ and s' are not real physical objects. The physical description of the spin state of the system is completely contained in the spinors $u(p,\, s)$ and $u(p',\, s')$. So no s^μ's appear in the matrix element, and the invariant functions f_i and g_i only depend on q^2, not $p \cdot s'$, *etc.* The spin polarization vectors s^μ and s'^μ are just convenient devices invented to describe the spinors.

Now let us use the fact that the strong interactions are invariant under time reversal invariance to show that f_i and g_i are all real. To simplify our lives, we will actually use invariance under PT. This is simpler because it involves changing the sign of all the coordinates, so we do not have to keep separate track of the time and space components of vectors. They all do the same thing.

Suppose we PT transform a matrix element (4.5.1):

$$\begin{aligned}
\langle B',\, p',\, s'|j^\mu_{(5)}|B,\, p,\, s\rangle^{PT} &= \pm\langle B,\, p,\, -s|j^{\mu\dagger}_{(5)}|B',\, p',\, -s'\rangle \\
&= \pm\langle B',\, p',\, -s'|j^\mu_{(5)}|B,\, p,\, -s\rangle^*
\end{aligned} \tag{4.5.4}$$

This needs some explanation. The PT is antiunitary, so it interchanges bras and kets and takes operators into their adjoints. The $+(-)$ sign for the vector (axial) current is not obvious, but we will come back to it. T changes the signs of momenta and spins, but then P changes the momenta back.

To see what PT does to spinor fields, consider the Dirac equation in a Majorana basis:

$$(i\partial\!\!\!/ - m)\psi(x) = 0. \tag{4.5.5}$$

We expect the PT–transformed field to be proportional to $\psi^*(-x)$. In this basis $*$ doesn't do much because the γ's are imaginary, but to change the sign of x. We also need something that anticommutes with the γ^μ's. This just γ_5, so

$$\psi(x)^{PT} \to \gamma_5\psi^*(-x). \tag{4.5.6}$$

The currents then transform as

$$j^\mu(0) = \overline{\psi}_1\gamma^\mu\psi_2 \to \psi_2^\dagger\,[\gamma^0\gamma^\mu]^T\psi_1 = \overline{\psi}_2\gamma^\mu\psi_1 = j^\mu(0)^* \tag{4.5.7}$$

and

$$j_5^\mu(0) = \overline{\psi}_1\gamma^\mu\gamma_5\psi_2 \to \psi_2^\dagger\,[\gamma^0\gamma^\mu\gamma_5]^T\psi_1 = -\overline{\psi}_2\gamma^\mu\gamma_5\psi_1 = j_5^\mu(0)^*. \tag{4.5.8}$$

Note that there is no $(-)$ sign due to Fermi statistics when we interchange 1 and 2, because this is done not by us, but by the antiunitary transformation that reverses all operator products.

Now the spinors describing the PT–transformed states are obtained similarly:

$$u(p,\,-s) = \gamma_5 u(p,\,s)^*. \tag{4.5.9}$$

Notice that

$$(p\!\!\!/ - m)u(p,\,-s) = 0 \tag{4.5.10}$$

and the γ_5 changes the spin because in anticommutes with $s\!\!\!/$. Putting all this together, we find (4.5.1)–(4.5.2), (4.5.4), and (4.5.7)–(4.5.9) yield the desired result that f_i and g_i are real.

4.6 Second Class Currents

We can use the results of Section 4.5 to eliminate two of the coefficients in the matrix elements (4.5.1)–(4.5.2), at least in the $SU(3)$ symmetry limit. First, let's consider the matrix element of a diagonal current

$$\begin{aligned} j^\mu_{(5)} &= \overline{\psi}\gamma^\mu(\gamma_5)\psi \\ \psi &= u,\ d,\ \text{or}\ s \end{aligned}\,, \tag{4.6.1}$$

with $B' - B$, since the currents do not change quark flavors. Note that both the vector and axial currents are Hermitian. Thus

$$\langle B,\,p',\,s'\,|j^\mu_{(5)}|B,\,p,\,s\rangle = \langle B,\,p,\,s\,|j^\mu_{(5)}|B,\,p',\,s'\rangle^*. \tag{4.6.2}$$

These can only be satisfied if

$$f_3 = g_2 = 0. \tag{4.6.3}$$

The other terms are all allowed, f_1 and f_2 obviously so, since they are just like the electromagnetic form factors.

Now for the off–diagonal currents, the situation is the same in the symmetry limit. For example, consider

$$\overline{u}\gamma^\mu(\gamma_5)d, \tag{4.6.4}$$

which is a linear combination of isospin currents. But the diagonal isospin current

$$j^\mu_{(5)3} = \overline{\psi}\gamma^\mu(\gamma_5)\frac{\tau_3}{2}\psi \tag{4.6.5}$$

has vanishing f_3 (g_2) by the preceding argument. If isospin is a good symmetry, (4.6.4) must also have f_3 and g_2 equal to zero. Similarly

$$\overline{d}\gamma^\mu(\gamma_5)s \tag{4.6.6}$$

gives vanishing f_3 (g_2) in the limit that the $SU(2)$ that mixes d and s (it is called U–spin) is a good symmetry. And

$$\overline{u}\gamma^\mu(\gamma_5)s \tag{4.6.7}$$

has f_3 $(g_2) = 0$ when the $SU(2)$ that mixes x and s (V–spin) is a good symmetry.

The f_3 and g_2 terms in the matrix elements of the off–diagonal currents are called "second–class currents" for obscure historical reasons.

4.7 The Goldberger-Treiman Relation

In the symmetry limit, the currents are conserved, so if we contract (4.5.1)–(4.5.2) with q_μ, we must get zero. For (4.5.1), current conservation is automatic if $f_3 = 0$, so it gives us nothing new. We already knew that $f_3 = 0$ in the symmetry limit. But (4.5.2) gives a peculiar relation

$$(m + m')g_1(q^2) = q^2 g_3(q^2). \tag{4.7.1}$$

It is peculiar because of the explicit power of q^2 on the RHS. Either g_1 goes to zero as $q^2 \to 0$ or $g_3(q^2)$ has a pole. To be specific, let the current be the axial current in neutron decay,

$$j^\mu_5 = \overline{u}\gamma^\mu\gamma_5 d. \tag{4.7.2}$$

We have repeatedly argued that because the u and d quark masses are very small compared to Λ, this current is approximately conserved. Thus, $g_1(0)$ in the symmetric world should be very close to $g_1(q^2)$ measured at very small momenta ($q^2 \simeq m_e^2$) in neutron decay in the real world. Then $g_1(0) \simeq -g_A \simeq 1.25$. We therefore believe that in the symmetric world, $g_3(q^2)$ has a pole. In fact, this is another signal of spontaneous symmetry breaking. The pole is due to the process shown in Figure 4–1.

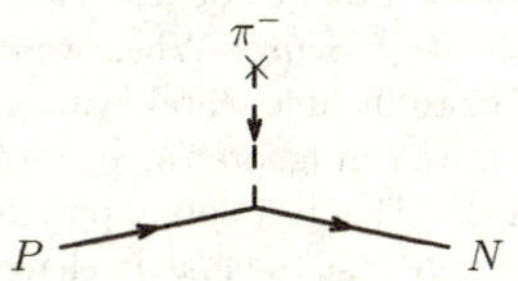

Figure 4-1:

Where $\times$ indicates the current annihilating the π^-. This diagram contributes

$$\overline{u}(p', s')\frac{q^\mu}{q^2} f_\pi \sqrt{2}\, g_{\pi NN} u(p, s) \tag{4.7.3}$$

to the matrix element of the axial vector current. So it contributes the pole term to $g_3(q^2)$. Comparing (4.7.1) and (4.7.3), we find

$$m_N g_A = \frac{f_\pi g_{\pi NN}}{\sqrt{2}} = F_\pi g_{\pi NN}. \tag{4.7.4}$$

This is the Goldberger-Treiman relation again, now in a more sophisticated version including strong–interactions effects in g_A, the renormalization of the axial vector current (compare (2.6.24)).

4.8 $SU(3)$ – D and F

Let us summarize what we know about the parameters in semileptonic baryon decay. The term $f_1(0)$ can be calculated from first principles similar to those used in Section 4.2. Furthermore, $f_1(q^2)$ and $f_2(q^2)$ can be related to the electromagnetic form factors in the $SU(3)$ symmetry limit. The principle here is not really different from the general discussion that follows, but it is simpler to say in this case, so we will illustrate it separately. The electromagnetic current can be written in terms of the $SU(3)$ currents as

$$j^\mu_{EM} = j^\mu_3 + \frac{1}{\sqrt{3}} j^\mu_8. \tag{4.8.1}$$

The j^μ_3 part is an isospin partner of the current we are actually interested in:

$$j^\mu = j^\mu_1 + i j^\mu_2 = \overline{u}\gamma^\mu d. \tag{4.8.2}$$

So we can use isospin considerations to show

$$\langle P|j^\mu|N\rangle = \langle P|j_{EM}|P\rangle - \langle N|j_{EM}|N\rangle. \tag{4.8.3}$$

Thus, the matrix element of interest in vector–current weak interactions is determined by the matrix element of the electromagnetic current. This relation was noticed long ago by Feynman and Gell-Mann. It was very important historically because it was a clue to the current–current structure of the weak interactions. In an analogous way we can use V–spin arguments to relate the matrix elements of the strangeness changing current $\overline{u}\gamma^\mu s$ to the electromagnetic current matrix elements.

Of the functions $g_i(q^2)$, we can safely ignore $g_2(q^2)$ because it vanishes in the $SU(3)$ symmetry limit. $g_3(q^2)$ can be calculated in terms of $g_1(q^2)$ in the chiral symmetry limit (4.7.3). However, we should not trust the pole at $q^2 = 0$. It is due to Goldstone–boson exchange, as shown by Figure 4–1. But in the real world, the lightest hadrons are the π and K, so we might expect the q^{-2} to be replaced by $(q^2 - m_\pi^2)^{-1}$. Because of the rapid q^2 dependence of the pole, there is a big difference between a pole at $q^2 = 0$ and a pole at $q^2 = m_\pi^4$. When we discuss nonlinear chiral Lagrangians in the next section, we will learn how to include chiral symmetry–breaking (and $SU(3)$ breaking) effects in a systematic way. For now, we can ignore the g_3 term completely, so long as we stick to decays with electrons in the final state. This is a good approximation because the contribution of the g_3 term is proportional to $q_\mu l^\mu$, where l^μ is the matrix element of the leptonic current in (4.2.6). But the divergence of the electron current is proportional to m_e, which is very small.

Finally, we can use $SU(3)$ symmetry to calculate $g_1(q^2)$ for all the baryons in terms of just two functions. To see this, notice that the currents and the baryon states in the matrix element

$$\langle B'|j^\mu|B\rangle \tag{4.8.4}$$

are all $SU(3)$ octets. There are only two ways to put three octets together in an $SU(3)$ invariant way.

The $SU(3)$ invariant couplings of octets are very easy to write down in a matrix notation. The quark triplet

$$\psi = \begin{pmatrix} u \\ d \\ s \end{pmatrix} \tag{4.8.5}$$

act like a column vector under $SU(3)$ transformations. The conjugate fields

$$\overline{\psi} = (\overline{u},\, \overline{d},\, \overline{s}) \tag{4.8.6}$$

act like a row vector. We can put these together to form a matrix
psi$\overline{\psi}$ that contains both $SU(3)$ singlet and octet pieces. We can get rid of the singlet part by making it traceless.

$$\psi\overline{\psi} - \tfrac{1}{3}I\,\mathrm{tr}\,(\psi\overline{\psi}) \equiv M =$$

$$\begin{pmatrix} (2u\overline{u} - d\overline{d} - s\overline{s})/3 & u\overline{d} & u\overline{s} \\ d\overline{u} & (-u\overline{u} + 2d\overline{d} - s\overline{s})/3 & d\overline{s} \\ s\overline{u} & s\overline{d} & (-u\overline{u} - d\overline{d} + 2s\overline{s})/3 \end{pmatrix}, \tag{4.8.7}$$

$$\propto \begin{pmatrix} \pi^0/\sqrt{2} + \eta/\sqrt{6} & \pi^+ & k^+ \\ \pi^- & -\pi^0/\sqrt{2} + \eta/\sqrt{6} & K^0 \\ K^- & \overline{K^0} & -2\eta/\sqrt{6} \end{pmatrix}$$

where we have identified the quark wave functions with the appropriate pseudoscalar mesons. Note that

$$\pi^0 = (u\overline{u} - d\overline{d})/\sqrt{2}, \quad \eta = (u\overline{u} - d\overline{d} - 2s\overline{s})/\sqrt{6}. \tag{4.8.8}$$

In (4.8.5)–(4.8.8) we are ignoring all Dirac indices and coordinate dependence; this matrix notation is just a way of keeping track of the $SU(3)$ properties.

M is a pseudoscalar meson wave function. We can find an octet baryon wave function in two ways. the simplest is just to replace each meson in (4.8.7) with the baryon with the same isospin and hypercharge: $\eta \to \Lambda$, $\pi^- \to \Sigma^-$, $K^+ \to P$, and so forth. Alternatively, we can construct quark wave functions by replacing $\overline{\psi}$ by

$$(ds - sd,\; su - us,\; ud - du), \tag{4.8.9}$$

which also behaves like a row vector. Either way, we find a baryon matrix

$$B = \begin{pmatrix} \Sigma^0/\sqrt{2} + \Lambda/\sqrt{6} & \Sigma^+ & P \\ \Sigma^- & -\Sigma^0/\sqrt{2} + \Lambda/\sqrt{6} & N \\ \Xi^- & \Xi^0 & -2\Lambda/\sqrt{6} \end{pmatrix}. \tag{4.8.10}$$

Note that the meson matrix is Hermitian in the sense that if we transpose it and charge conjugate, we get the same states back again. But the baryon matrix is not Hermitian because charge conjugation does not give back the same states; it changes baryons to antibaryons.

To build an $SU(3)$ invariant out of these matrices, we just saturate all the indices by taking traces of products. For example, the $SU(3)$ invariant meson–baryon coupling

$$\langle B'|MB\rangle \tag{4.8.11}$$

is proportional to the matrices M and B and to the Hermitian conjugate matrix B' (because B' appears in a bra). There are two different ways of combining these three matrices to form an $SU(3)$ singlet:

$$\text{tr } MB\overline{B'} \quad \text{or} \quad \text{tr } M\overline{B'}B.(4.8.12) \tag{4.8.12}$$

These are conventionally combined into combinations involving the commutator and anticommutator of two of the matrices and called F and D, so that

$$\langle B'|MB\rangle = D\text{tr}\left(M\{B,\overline{B'}\}\right) + F\text{tr}\left(M[B,\overline{B'}]\right), \tag{4.8.13}$$

up to conventional normalization. The reason for the names is that the F coupling is proportional to the structure constant f_{abc} while the D coupling is proportional to the symmetric invariant tensor $d_{abc} = 2\text{tr }(T_a\{T_b, T_c\})$.

Similarly, if we write a current in an $SU(3)$ notation in terms of a traceless 3×3 matrix T such that

$$\gamma^\mu_{(5)} = \overline{\psi}\gamma^\mu(\gamma_5)T\psi, \tag{4.8.14}$$

then the $SU(3)$ invariant matrix element is

$$\langle B'|j_5^\mu|B\rangle = D\text{tr}\left(T\{B,\overline{B'}\}\right) + F\text{tr}\left(T[B,\overline{B'}]\right). \tag{4.8.15}$$

All the spinors, momentum dependence, and the rest of the matrix elements like (4.5.1)–(4.5.2) are contained in D and F. All the particle and current labels are in the traces. Thus, this enables us to relate the invariants (for example, g_1) in one matrix element to those in another.

Let us illustrate its use by explicity writing out (4.8.5) for the current of interest in the charged–current weak interactions with

$$T = \begin{pmatrix} 0 & V_{ud} & V_{us} \\ 0 & 0 & 0 \\ 0 & 0 & 0 \end{pmatrix}. \tag{4.8.16}$$

If we explicitly evaluate (4.8.15) using (4.8.10) and its Hermitian conjugate $\overline{B}$ (we will drop the prime because $\overline{B'}$ always appears barred because it is in a bra), we get

$$\begin{aligned}
&F\Big\{V_{ud}\left[\sqrt{2}\overline{\Sigma}^0\Sigma^- - \sqrt{2}\overline{\Sigma}^+\Sigma^0 + \overline{P}N - \overline{\Xi}^0\Xi^-\right] \\
&+V_{us}\Big[\sqrt{1/2}\overline{\Sigma}^0\Xi^- + \sqrt{3/2}\overline{\Lambda}\Xi^- + \overline{\Sigma}^+\Xi^0 \\
&-\sqrt{1/2}\overline{\Sigma}^0\Xi^- - \sqrt{3/2}\overline{P}\Lambda - \overline{N}\Sigma^-\Big]\Big\} \\
&+D\Big\{V_{ud}\left[\sqrt{2/3}\overline{\Sigma}^+\Lambda + \overline{P}N + \overline{\Xi}^0\Xi^-\right] \\
&+V_{us}\Big[\sqrt{1/2}\overline{\Sigma}^0\Xi^- - \sqrt{1/6}\overline{\Lambda}\Xi^- \\
&+\sqrt{1/2}\overline{P}\Sigma^0 - \sqrt{1/6}\overline{P}\Lambda + \overline{N}\Sigma^-\Big]\Big\}
\end{aligned} \tag{4.8.17}$$

Notice the F term for the V_{ud} current is just an isospin raising–operator matrix element.

As we have discussed above, the f_1 and f_2 terms are very well constrained because they are related to the electromagnetic current associated with

$$T = \begin{pmatrix} \frac{2}{3} & 0 & 0 \\ 0 & -\frac{1}{3} & 0 \\ 0 & 0 & -\frac{1}{3} \end{pmatrix} . \tag{4.8.18}$$

The axial vector current is not so well determined. One combination that is very well measured is g_A, which appears in neutron decay. Thus

$$V_{ud}(D+F) = g_{1_{N\to P}} = -g_A \simeq 1.25. \tag{4.8.19}$$

Only the ratio D/F for the axial vector current remains to be determined. This can be fit to all the semileptonic baryon decay rates. This also, of course, determined the KM parameter V_{us}. This is where the value in your particle data book comes from.

$$V_{us} = 0.218 \text{ to } 0.224 \tag{4.8.20}$$

Problems

4-1. Calculate the rate for the decay $K^+ \to \pi^0 e^+ \nu_e$ in terms of G_F and V_{us}. Assume the $SU(3)$ value for the hadronic matrix element, and set $m_\pi = m_e = 0$ in your calculation.

4-2. Calculate the rate for $\pi^+ \to \mu^+ \nu_\mu$ and $\pi^+ \to e^+ \nu_e$ in terms of f_π, G_F, and the lepton masses.

4-3. The energy-momentum tensor, $T^{\mu\nu}(x)$, is symmetric, conserved, and related to the 4-momentum operators by

$$P^\mu = \int T^{\mu 0}(x)\, d^3x$$

Use arguments analogous to (4.2.8-4.2.17) to calculate the matrix elements of $T^{\mu\nu}$ between spinless particle states.

4-4. Show that (4.5.1) and (4.5.2) are the most general matrix elements consistent with Lorentz invariance and parity invariance of the strong interactions.

4-5. Derive (4.8.3). Give an example of a similar relation involving the strangeness-changing current $\overline{u}\gamma^\mu s$.

5 — Chiral Lagrangians — Goldstone Bosons

5.1 $SU(3) \times SU(3) \to SU(3)$

In Section 2.6 we discussed the σ–model of nucleons and pions with spontaneously broken $SU(2) \times SU(2)$ chiral symmetry. In an obvious way, we could recast the model into a more modern language by replacing the nucleon doublet (P, N) by the light quark doublet (u, d). To describe the strange quark, we must extend the model to encompass an $SU(3) \times SU(3)$ chiral symmetry. The obvious way to do this is simply to replace the 2×2 σ field by a 3×3 Σ field, transforming like

$$\Sigma \to L \Sigma R^{\dagger} = \Sigma' \tag{5.1.1}$$

under independent $SU(3)$ transformations of the left- and right-handed quark triplets by the unitary unimodular matrices L and R.

However, it turns out that the 2×2 case is rather special. In the 3×3 case, we cannot restrict the field as we did in (2.6.11). We can require that $\det \Sigma$ is real. But if the components of Σ are to transform linearly into one another under (5.1.1), Σ must be a general 3×3 matrix, except for the reality of the determinant. Thus it depends on 17 real parameters. The reason is that the 3–dimensional representation of $SU(3)$ is complex. There is no matrix S that satisfies $-ST_a^* S^{-1} = T_a$ for the $SU(3)$ generators T_a.

We suspect from our experience with the σ–model and f_π that the $SU(3) \times SU(3)$ symmetry will have to be spontaneously broken if this model is to have anything to do with our world. There is now a tremendous amount of phenomenological and theoretical evidence that the symmetry is spontaneously broken down to $SU(3)$ with a VEV

$$\langle \Sigma \rangle = \begin{pmatrix} f & 0 & 0 \\ 0 & f & 0 \\ 0 & 0 & f \end{pmatrix} . \tag{5.1.2}$$

For (5.1.2), since there is a broken $SU(3)$ symmetry, all the ordinary $SU(3)$ generators annihilate the vacuum. The chiral $SU(3)$ generators are associated with Goldstone bosons. The Goldstone bosons are therefore an $SU(3)$ octet of pseudoscalars. There are good candidates for all of these, and indeed they are the lightest meson states, the pseudoscalar octet, π, K, $\overline{K}$, and η (see Table 3.1).

5.2 Effective Low–Momentum Field Theories

We can easily build a renormalizable theory based on the 3×3 Σ field in which the field develops a VEV of the form (5.1.2) and spontaneously breaks the the $SU(3) \times SU(3)$ symmetry down to

$SU(3)$. But we have no particular reason to believe that has anything to do with QCD. These two field theories share a spontaneously broken $SU(3) \times SU(3)$ symmetry (in the limit of QCD in which $m_\mu = m_d = m_s = 0$), and thus, as we will discuss in more detail, they both contain an $SU(3)$ octet of Goldstone bosons. But they have little else in common. The key fact about Goldstone bosons is that they are massless because they are related to the vacuum state by the chiral symmetries. We expect QCD to give such Goldstone bosons just because the symmetry is spontaneously broken, but until we get smarter, we can't say much about them that depends on the details of the dynamics. So we would like to learn everything we can about the Goldstone bosons without assuming anything but the spontaneously broken chiral symmetries.

A chiral charge (if such a thing actually made sense) would act on the vacuum state to produce a zero–momentum Goldstone boson. This is why the charge doesn't exist — the particle state and vacuum state are normalized differently. Thus, we would expect the chiral symmetries to relate processes involving different numbers of zero–energy Goldstone bosons. We want, therefore, to study Goldstone bosons at low energies. In fact, at sufficiently low energies, only the Goldstone bosons can be produced because they are the lightest particles around, so we can start by studying a theory involving only Goldstone–boson fields.

We know that a QCD theory with three light quarks has a spontaneously broken $SU(3)\times SU(3)$ symmetry, but that is all we can know without understanding the dynamics in detail. Therefore, to incorporate the information about the symmetry without making unwarranted assumptions about the dynamics, we would like to build the most general low–energy theory we can, consistent with $SU(3) \times SU(3)$ symmetry. We will do this with two basic ideas: we will use a nonlinear representation of the chiral $SU(3)$ symmetry, and we will expand the Lagrangian in powers of the momentum — in order to concentrate on low–momentum physics.

Consider an exponential function of the eight Goldstone–boson fields π_a, the unitary, unimodular matrix

$$
\begin{aligned}
U &= \exp[2i\underset{\sim}{\Pi}/f] \\
\underset{\sim}{\Pi} &= \pi_a T_a,
\end{aligned}
\tag{5.2.1}
$$

where f is a constant with dimensions of mass. We will require that U transform linearly under $SU(3) \times SU(3)$:

$$
U \to U' = LUR^\dagger,
\tag{5.2.2}
$$

just as in (5.2.1). But now the Goldstone–boson fields transform nonlinearly:

$$
\pi_a \to \pi'_a, \qquad \underset{\sim}{\Pi} = \pi_a T_a \to \underset{\sim}{\Pi}' = \pi'_a T_a
\tag{5.2.3}
$$

where

$$
U' = \exp[2i\underset{\sim}{\Pi}'/f].
\tag{5.2.4}
$$

(5.2.2)–(5.2.4) define π'_a as a complicated nonlinear function of the π's and L and R.

To see what this looks like in detail, let us parametrize L and R as follows:

$$
\begin{aligned}
L &= e^{i\underset{\sim}{c}}\, e^{i\underset{\sim}{\epsilon}} \\
R &= e^{-i\underset{\sim}{c}}\, e^{i\underset{\sim}{\epsilon}},
\end{aligned}
\tag{5.2.5}
$$

where ϵ_a and c_a are real parameters. This parametrization is entirely general. We can calculate c_a by noting

$$LR^\dagger = e^{2i\underset{\sim}{c}}, \tag{5.2.6}$$

then ϵ_a can be calculated from (5.2.5).

If $c_a = 0$, the transformation is an ordinary $SU(3)$ transformation

$$U \to e^{i\underset{\sim}{\epsilon}}\, U e^{-i\underset{\sim}{\epsilon}}, \tag{5.2.7}$$

under which the π's transform *linearly*

$$\underset{\sim}{\Pi} \to e^{i\underset{\sim}{\epsilon}}\, \underset{\sim}{\Pi} e^{-i\underset{\sim}{\epsilon}}. \tag{5.2.8}$$

Thus ϵ_a describes the ordinary $SU(3)$ subgroup of $SU(3) \times SU(3)$. The Goldstone bosons are just an octet under the $SU(3)$ subgroup as they should be.

If $\underset{\sim}{\epsilon} = 0$, (5.2.2) is a pure chiral transformation.

$$U' = e^{i\underset{\sim}{c}}\, U e^{i\underset{\sim}{c}}. \tag{5.2.9}$$

To get a feeling for the meaning of (5.2.9), let us take c to be infinitesimal and write Π' as a power series in c and Π.

$$(1 + 2i\underset{\sim}{\Pi}'/f + \cdots) = (1 + i\underset{\sim}{c} + \cdots)(1 + 2i\underset{\sim}{\Pi}/f + \cdots)(1 + i\underset{\sim}{c} + \cdots). \tag{5.2.10}$$

Comparing the two sides, we see

$$\underset{\sim}{\pi}' = \underset{\sim}{\pi} + f\,\underset{\sim}{c} + \cdots \qquad \text{or} \qquad \pi_a' = \pi_a + f c_a + \cdots\,. \tag{5.2.11}$$

All the terms in the (5.2.11) are odd in π and c because of parity invariance. Both c and π change signs if we interchange L and R. The inhomogeneous term in (5.2.11) is a signal of spontaneous symmetry breakdown, as we will see later.

This special exponential form for the nonlinear representation is the simplest of an infinite number of possibilities. All the others are equivalent, however, in the sense that theories constructed out of them will have the same S–matrix as the theories we will write down. This is discussed (along with other interesting things) in an elegant pair of papers by S. Coleman *et al.* *Phys.Rev.* **177**:2239–2247 and 2247–2250, 1969).

The idea is as follows. Suppose we build the most general Lagrangian involving the Π's (or, equivalently, U) that is invariant under $SU(3) \times SU(3)$ (5.2.2)–(5.2.4). Since all we have used is the symmetry, our theory should be equivalent to a theory based on any other realization of the $SU(3) \times SU(3)$ symmetry on the Goldstone bosons. We use nonlinear realizations of $SU(3) \times SU(3)$ because we can then describe only the Goldstone–boson fields, not any extraneous fields introduced by a linear realization.

The Lagrangians we build will be complicated nonlinear functions of the Goldstone–boson fields. They will not be renormalizable. That does not concern us. They are designed to describe the low–energy behavior of the theory. Renormalizability has to do with the high–energy behavior.

We are going to organize the Lagrangian in terms of increasing powers of momentum or, equivalently, in terms of increasing numbers of derivatives. Thus, we look first at terms with no derivatives, then two, then four, and so on.

The terms with no derivatives are easy. There aren't any. Every invariant function of U without derivatives is just a constant. This will probably be obvious if you try to make one. Alternatively, you can note that the terms with no derivatives actually have a local $SU(3) \times SU(3)$ symmetry, which is enough to transform away from the Goldstone bosons completely (as in the Higgs mechanism). Thus, the non-derivative terms can have no dependence at all on the Goldstone–boson fields.

There is only one term with two derivatives. It is

$$\frac{f^2}{4}\text{tr}\left(\partial^\mu U^\dagger \partial_\mu U\right). \tag{5.2.12}$$

It is easy to see that all the other two derivative terms can be massaged into this same form. We have chosen the constant in front of (5.2.12) so that it contains the conventionally normalized π_a kinetic energy,

$$\frac{1}{2}\partial_\mu \pi_a \partial^\mu \pi_a. \tag{5.2.13}$$

The other terms in (5.2.12) describe the most important self–interactions of the π's. They are interesting but have little to do with weak interactions.

5.3 Sources

To discuss symmetry breaking and the currents in the low energy theory describing the Goldstone bosons, it is convenient to start with the massless QCD theory for the triplet q of light quarks, with the explicit $SU(3) \times SU(3)$ symmetry, and add classical sources in the following way

$$\begin{aligned}\mathcal{L} &= \mathcal{L}^0_{\text{QCD}} + \overline{q}\gamma^\mu(v_\mu + a_\mu\gamma_5)q - \overline{q}(s + ip\gamma_5)q \\ &= \mathcal{L}^0_{\text{QCD}} + \overline{q}_L\gamma^\mu \ell_\mu q_L + \overline{q}_R\gamma^\mu r_\mu q_R - \overline{q}_R(s+ip)q_L - \overline{q}_L(s-ip)q_R\end{aligned} \tag{5.3.1}$$

where $\mathcal{L}^0_{\text{QCD}}$ is the massless QCD Lagrangian and v_μ, a_μ, r_μ, ℓ_μ, s and p are classical Hermitian 3×3 matrix fields (we can take the vector fields to be traceless), and

$$r_\mu = v_\mu - a_\mu\,, \qquad \ell_\mu = v_\mu + a_\mu\,. \tag{5.3.2}$$

The r_μ and ℓ_μ fields are classical gauge fields for the $SU(3)_R$ and $SU(3)_L$ symmetries, and also the sources for the corresponding Noether currents.

The theory described by (5.3.1) is invariant under an $SU(3) \times SU(3)$ gauge symmetry under which the fields and sources transform as follows:

$$\begin{aligned} q_R &\to R\,q_R \\ q_L &\to L\,q_L \\ r_\mu &\to R\,r_\mu\,R^\dagger + iR\,\partial_\mu\,R^\dagger \\ \ell_\mu &\to L\,\ell_\mu\,L^\dagger + iL\,\partial_\mu\,L^\dagger \\ s+ip &\to R\,(s+ip)\,L^\dagger \end{aligned} \tag{5.3.3}$$

where R and L are unitary 3×3 matrices describing the $SU(3)_R$ and $SU(3)_L$ symmetries.

Thus we want to build a low energy theory with the same symmetries —

$$
\begin{aligned}
U &\to L\,U\,R^\dagger \\
r_\mu &\to R\,r_\mu\,R^\dagger + iR\,\partial_\mu\,R^\dagger \\
\ell_\mu &\to L\,\ell_\mu\,L^\dagger + iL\,\partial_\mu\,L^\dagger \\
s + ip &\to R\,(s+ip)\,L^\dagger
\end{aligned}
\tag{5.3.4}
$$

The low energy theory will be described by an effective Lagrangian that depends on U and the classical sources,

$$\mathcal{L}(U, r, \ell, s, p) \tag{5.3.5}$$

As usual, we want to build the most general such effective Lagrangian consistent with the symmetries, (5.3.4). This is only useful because the sources, like derivatives, can be thought of as small. Thus we can expand (5.3.5) in powers of the sources and truncate the expansion after a small number of terms.

Because the symmetry, (5.3.4), is a gauge symmetry, the derivatives in (5.2.12) must be promoted to covariant derivatives,

$$\frac{f^2}{4}\mathrm{tr}\left(D^\mu U^\dagger D_\mu U\right). \tag{5.3.6}$$

where

$$D^\mu U = \partial^\mu U - i\ell^\mu U + iUr^\mu \qquad D^\mu U^\dagger = \partial^\mu U - ir^\mu U + iU\ell^\mu \tag{5.3.7}$$

Note also that the QCD theory is invariant under a parity transformation, which on the fields of the low energy theory takes the form

$$
\begin{aligned}
U &\leftrightarrow U^\dagger \\
\ell^\mu &\leftrightarrow r^\mu \\
s &\to s \\
p &\to -p
\end{aligned}
\tag{5.3.8}
$$

along with the parity transformation in space, $\vec{r} \to -\vec{r}$.

5.4 Symmetry breaking and light quark masses

We do not want massless Goldstone bosons. We expect the masses of the observed pseudoscalar octet to be due to the $SU(3) \times SU(3)$ symmetry breaking. In QCD, we know precisely what breaks the symmetry. It is the quark term

$$\bar{q}_L M q_R + \text{h.c.}, \tag{5.4.1}$$

where M is the quark mass matrix

$$M = \begin{pmatrix} m_u & 0 & 0 \\ 0 & m_d & 0 \\ 0 & 0 & m_s \end{pmatrix} \tag{5.4.2}$$

But it is easy to incorporate the quark mass matrix in the theory with sources, with the symmetries of (5.3.4). We simply have to build a low energy theory with the right symmetries, and then set $s = M$. Of course, M is a constant and doesn't transform under $SU(3) \times SU(3)$, so setting $s = M$ actually breaks the symmetry. By itself, this is not very useful because with an arbitrary function of M, we can break the symmetry in an arbitrarily complicated way. To extract information, we assume that we can expand (5.3.5) in powers of s and p (which always appear in the combination $s + ip$ or its adjoint, which transform simply under (5.3.4)) and truncate the expansion assuming that the symmetry breaking is small. We expect this to be an excellent approximation for the u and d quark masses, which seem to be very small compared to the QCD scale parameter Λ. It is plausible, but not obviously justified for m_s. We will see how it works.

To first approximation, we want to find the invariant function of U and $s \pm ip$ that is linear in $s \pm ip$. This is unique. It is

$$v^3 \mathrm{tr}\left[U(s+ip)\right] + \mathrm{h.c.} \tag{5.4.3}$$

To see what this means, we again expand in powers of Π, and we set $s = M$ and $p = 0$. If v and M are real, the linear term cancels and the quadratic term is

$$-4\frac{v^3}{f^2}\mathrm{tr}\,(M\underset{\sim}{\Pi}^2), \tag{5.4.4}$$

which corresponds to a mass term

$$8\frac{v^3}{f^2}\mathrm{tr}\,(M\underset{\sim}{\Pi}^2) \tag{5.4.5}$$

for the pseudoscalar mesons.

We will first evaluate these masses in the limit of isospin invariance, ignoring weak and electromagnetic interactions and setting $m_u = m_d = m$. Then

$$\begin{aligned}
m_\pi^2 &= 8\tfrac{v^3}{f^2}\mathrm{tr}\left(M\begin{pmatrix}\frac{1}{4} & 0 & 0\\ 0 & \frac{1}{4} & 0\\ 0 & 0 & 0\end{pmatrix}\right) = 4\tfrac{v^3}{f^2}m\\
m_K^2 &= 8\tfrac{v^3}{f^2}\mathrm{tr}\left(M\begin{pmatrix}\frac{1}{4} & 0 & 0\\ 0 & 0 & 0\\ 0 & 0 & \frac{1}{4}\end{pmatrix}\right) = 2\tfrac{v^3}{f^2}(m+m_s)\\
m_\eta^2 &= 8\tfrac{v^3}{f^2}\mathrm{tr}\left(M\begin{pmatrix}\frac{1}{12} & 0 & 0\\ 0 & \frac{1}{12} & 0\\ 0 & 0 & \frac{1}{3}\end{pmatrix}\right) = 4\tfrac{v^3}{f^2}\left(\tfrac{m}{3}+2\tfrac{m_s}{3}\right)
\end{aligned} \tag{5.4.6}$$

The m^2 determined in this way satisfy the Gell-Mann Okubo relation

$$3m_\eta^2 + m_\pi^2 = 4m_K^2. \tag{5.4.7}$$

The interesting thing about this is that we have derived the GMO relation for m^2 specifically. In an ordinary $SU(3)$ symmetry argument, the relation should apply equally well for linear masses. In fact, the GMO us much better satisfied for m^2 than for m for the pseudoscalars. The only coherent explanations of this fact are equivalent to the derivation we have just given. So this is some evidence for the validity of the nonlinear chiral theory. Notice that the approximation we have used of keeping two derivative terms in the invariant part of $\mathcal{L}$ and the term linear in M in the symmetry breaking part makes a kind of sense. As (5.4.5) shows, if the pseudoscalar boson momentum is of the order of its mass, the two terms are of the same order of magnitude.

Note also that (5.4.6) does not determine the scale of the quark masses. If we define the parameter μ by

$$\mu = 2v^3/f^2, \tag{5.4.8}$$

then the pseudoscalar meson mass squares are proportional to μM. In the expansion in powers of momenta and quark masses, μM counts as two powers of momentum or two derivatives.

To this order, the effective Lagrangian has the form

$$\mathcal{L} = f^2 \left\{ \frac{1}{4} \text{tr} \left(D^\mu U^\dagger D_\mu U \right) + \frac{1}{2} \text{tr} \left[U^\dagger \mu (s - ip) \right] + \frac{1}{2} \text{tr} \left[U \mu (s + ip) \right] \right\}. \tag{5.4.9}$$

5.5 What Happened to the Axial $U(1)$?

If something is bothering you about this treatment of Goldstone bosons from QCD, you are to be commended. The free massless quark theory has a chiral $SU(3) \times U(1) \times SU(3) \times U(1)$. Why don't we include a Goldstone boson corresponding to a spontaneously breaking chiral $U(1)$? This would mean adding a ninth π, π_0 corresponding to a $U(1)$ generator

$$T_0 = \sqrt{1/6} \begin{pmatrix} 1 & 0 & 0 \\ 0 & 1 & 0 \\ 0 & 0 & 1 \end{pmatrix} \tag{5.5.1}$$

It is not hard to see that this is a phenomenological disaster. There is a ninth pseudoscalar, the η' at 958 MeV. It seems to be an $SU(3)$ singlet, primarily. But if we followed the analysis of Section 5.3, we would find something very different. Instead of an $SU(3)$ octet and singlet like the η and η', there would be an ideally mixed pair like the ω *and* ϕ in the vector meson system, an isoscalar degenerate with the pion and another consisting of almost pure $s\bar{s}$ with mass proportion to m_s. Even if the f constant is allowed to be different for the π_0, the mass spectrum of the almost Goldstone bosons does not look like what we see for the pseudoscalar mesons (see Problem 5–1).

This was a serious problem for many years until it was discovered that the axial $U(1)$ is actually not a symmetry of the QCD theory. The symmetry breakdown can be traced to the appearance of a peculiar gauge invariant interaction term,

$$\frac{\theta}{64\pi^2} g^2 \epsilon_{\mu\nu\lambda\sigma} G_a^{\mu\nu} G_a^{\lambda\sigma}. \tag{5.5.2}$$

This term had always been neglected because it had been a total divergence. But using semiclassical "instanton" techniques, 't Hooft showed that it could not be neglected, essentially because the gauge fields relevant in the quantum theory do not fall off fast enough at infinity to allow neglect of surface terms.

This is related to the axial $U(1)$ problem because of yet another peculiar effect, the anomaly. The gauge–invariant axial current

$$j_5^\mu = \bar{q} \gamma^\mu \gamma_5 T q \tag{5.5.3}$$

in the massless quark QCD theory is not conserved unless $\text{tr}\, T = 0$, even though canonical manipulations would lead us to believe it is. The problem is that the triangle diagrams in which the current couples to two gluons through a quark loop are linear divergent. So the canonical reasoning that leads to $\partial_\mu j_5^\mu = 0$ breaks down and we find instead that

$$\partial_\mu j_5^\mu = \text{tr}\, T \frac{g^2}{32\pi^2} \epsilon_{\mu\nu\lambda\sigma} G_a^{\mu\nu} G_a^{\lambda\sigma}. \tag{5.5.4}$$

Because the divergence of the current involves the term (5.5.2), the associated symmetry does not leave the Lagrangian invariant, it changes θ (this follows from our Noether's theorem argument [1.4.6]).

This doesn't effect the $SU(3)$ currents because tr $T = 0$, but it eliminates the axial $U(1)$ as a symmetry! Thus the analysis of the previous sections, in which do not include the π_0 as a Goldstone boson, is all right.

5.6 Light Quark Mass Ratios

To extract reliable values for the ratios of m_u, m_d, and m_3 to one another, we must include a treatment of isospin breaking. There are two important sources of isospin breaking, the u — d mass difference and the electromagnetic interactions. We will take them one at a time.

If we perform the analysis of Section 5.3 without assuming $m_u = m_d$, we find

$$\begin{aligned} m_{\pi^\pm}^2 &= \mu(m_u + m_d) \\ m_{K^\pm}^2 &= \mu(m_u + m_s) \\ m_{K^0}^2 &= \mu(m_d + m_s) \end{aligned} \tag{5.6.1}$$

And for the π_s, π_8 system, we find a 2×2 mass–squared matrix:

$$\mu \begin{pmatrix} m_u + m_d & (m_u - m_d)/\sqrt{3} \\ (m_u - m_d)/\sqrt{3} & (4m_s + m_u + m_d)/3 \end{pmatrix}. \tag{5.6.2}$$

The off–diagonal terms in (5.6.2) produce first–order mixing of the isovector and isoscalar states, but their effect on the eigenvalues is second order in isospin breaking and very small. So we will ignore it and write

$$\begin{aligned} m_{\pi^0}^2 &\simeq \mu(m_u + m_d) \\ m_\eta^2 &\simeq \tfrac{\mu}{3}(4m_s + m_u + m_d) \end{aligned} \tag{5.6.3}$$

The easiest way to understand the effect of the electromagnetic interactions is to think of them as an additional chiral symmetry–breaking effect. Because the electric charge matrix

$$Q = \begin{pmatrix} \frac{2}{3} & 0 & 0 \\ 0 & -\frac{1}{3} & 0 \\ 0 & 0 & -\frac{1}{3} \end{pmatrix} \tag{5.6.4}$$

commutes with the chiral $SU(3)$ generators associated with π^0, K^0, $\overline{K^0}$, and η fields, the EM interactions do not break the associated chiral symmetries. Thus, if there were no quark mass terms, the π^0, K^0, $\overline{K^0}$, and η would remain as exactly massless Goldstone bosons. They would get no contribution to their masses from electromagnetism. In the spirit of chiral perturbation theory, we will assume that this situation persists approximately in the presence of quark masses. This amounts to ignoring contributions that are suppressed both by quark masses and α.

The chiral symmetries associated with the $K^\pm$ and $\pi^\pm$ fields are broken by the electromagnetic interactions, so that the $K^\pm$ and $\pi^\pm$ do get a mass contribution from electromagnetism. But because the d and s charges are the same, the $K^\pm$ and $\pi^\pm$ get the same contribution (at least in the limit that the quark masses vanish). The arguments of the last two paragraphs are part of an analysis called Dashen's theorem. It enables us to parametrize the effect of the EM interactions in terms of a single number, the contribution to the $\pi^\pm$ mass–squared, which will call Δm^2. Thus, we can write all the masses as follows:

$$\begin{aligned} m_{\pi^\pm}^2 &= \mu(m_u + m_d) + \Delta m^2 \\ m_{K^\pm}^2 &= \mu(m_u + m_s) + \Delta m^2 \\ m_{K^0}^2 &= \mu(m_d + m_s) \\ m_{\pi^0}^2 &= \mu(m_u + m_d) \\ m_{\eta^0}^2 &= \mu(4m_s + m_u + m_d)/3. \end{aligned} \tag{5.6.5}$$

Putting in observed π and K masses, we get

$$\begin{aligned} \mu m_u &= 0.00649\text{GeV}^2 \\ \mu m_d &= 0.01172\text{GeV}^2 \\ \mu m_s &= 0.23596\text{GeV}^2. \end{aligned} \tag{5.6.6}$$

This gives the ratios

$$\frac{m_s}{m_d} = 20, \qquad \frac{m_u}{m_d} = 0.55\ . \tag{5.6.7}$$

Putting (5.6.6) into (5.6.5) gives an η mass of 566 MeV compared to the observed 549, about 3% off, but not bad.

5.7 The Chiral Currents

In this section, we derive and discuss the currents associated with chiral $SU(3) \times SU(3)$ transformations on our effective Lagrangian. The currents can be obtained easily by taking the functional derivative with respect to the classical gauge fields, as in (1b.1.13). If we are interested in matrix elements of a single current, we can then set the classical gauge fields to zero. Using (5.3.6) and (5.3.7), with

$$\ell^\mu = \ell_a^\mu T_a \qquad\qquad r^\mu = r_a^\mu T_a \tag{5.7.1}$$

we find for the left-handed current (which is involved in the semi-leptonic decays)

$$j_{La}^\mu = \frac{\delta \mathcal{L}}{\delta \ell_{a\mu}} = i\frac{f^2}{4}\,\text{tr}\left[U^\dagger T_a \partial^\mu U - \partial^\mu U^\dagger T_a U\right]. \tag{5.7.2}$$

But since $U^\dagger U = 1$, it follows that

$$(\partial^\mu U)U^\dagger = -U\partial^\mu U^\dagger, \tag{5.7.3}$$

so we can rewrite (5.7.2) as

$$j_{La}^\mu = i\frac{f^2}{2}\,\text{tr}\,(U^\dagger T_a \partial^\mu U). \tag{5.7.4}$$

The right-handed current can be easily obtained by the parity transformation, (5.3.8), which in this instance means interchanging U and $U^\dagger$.

We now want to evaluate (5.7.4) as a power series in Π. We can do it easily by means of a marvelous identity for the derivative of the exponential of a matrix (see Problem 5–2)

$$\partial^\mu e^M = \int_0^1 e^{sM}\,\partial^\mu M e^{(1-s)M}\,ds. \tag{5.7.5}$$

I love this identity because it is so easy to guess, yet rather nontrivial. At any rate, using (5.7.5), we find

$$j^\mu_{La} = -f\int_0^1 \text{tr}\left(T_a\cdot e^{2is\underset{\sim}{\Pi}/f}\,\partial^\mu\underset{\sim}{\Pi}e^{-2is\underset{\sim}{\Pi}/f}\right)ds\,. \tag{5.7.6}$$

It is now easy to expand in powers of $\underset{\sim}{\Pi}$:

$$\begin{aligned} j^\mu_{La} = \; & -f\text{tr}\left(T_a\partial^\mu\underset{\sim}{\Pi}\right) \\ & -i\text{tr}\left(T_a[\underset{\sim}{\Pi},\,\partial^\mu\underset{\sim}{\Pi}]\right) \\ & +\tfrac{2}{3f}\text{tr}\left(T_a\left[\underset{\sim}{\Pi},\,[\underset{\sim}{\Pi}\partial^\mu\underset{\sim}{\Pi}]\right]\right) \\ & +\cdots\,. \end{aligned} \tag{5.7.7}$$

These terms alternate. The terms odd in $\underset{\sim}{\Pi}$ are axial vectors while those even in $\underset{\sim}{\Pi}$ are vectors. The right-handed current is obtained trivially by switching the signs of all the odd terms.

In the quark theory, the corresponding current is

$$j^\mu_{La} = \overline{q}T_a\gamma^\mu\frac{(1+\gamma_5)}{2}q. \tag{5.7.8}$$

If we compare (5.7.7)–(5.7.8), we see that the axial isospin current is

$$j^\mu_{5a} = -f\partial^\mu\pi_a + \cdots, \tag{5.7.9}$$

where $a = 1$ to 3. Thus, from (4.3.6)

$$f = F_\pi = 93\text{ MeV}. \tag{5.7.10}$$

5.8 Semileptonic K Decays

The semileptonic decay amplitudes for $K \to \mu\nu,\ e\nu,\ \mu\nu\pi,\ e\nu\pi,\ e\nu\pi\pi$, and so on, are all proportional to matrix elements of the strangeness–changing charged current

$$\overline{u}\gamma^\mu(1+\gamma_5)s. \tag{5.8.1}$$

In terms of $SU(3)\times SU(3)$ currents,

$$\begin{aligned} j^\mu_a &= \overline{q}\gamma^\mu T_a q \\ j^\mu_{5a} &= \overline{q}\gamma^\mu\gamma_5 T_a q. \end{aligned} \tag{5.8.2}$$

This is

$$\begin{aligned}\overline{u}\gamma^\mu(1+\gamma_5)s &= \overline{q}\gamma^\mu(1+\gamma_5)(T_4+iT_5)q \\ &= j_4^\mu + ij_5^\mu + j_{54}^\mu + ij_{55}^\mu,\end{aligned} \tag{5.8.3}$$

because

$$T_4 + iT_5 = \begin{pmatrix} 0 & 0 & 1 \\ 0 & 0 & 0 \\ 0 & 0 & 0 \end{pmatrix}. \tag{5.8.4}$$

The currents associated with transformations on the left-handed quarks contain the projection operator $(1+\gamma_5)/2$, so they are

$$\overline{q}\gamma^\mu \frac{1+\gamma_5}{2} T_a q. \tag{5.8.5}$$

These are the currents that correspond to the j_L^μ in our effective theory. This correspondence depends only on the symmetry, and it determines the normalization as well as the form of the currents. Thus

$$j_{La}^\mu \leftrightarrow \frac{1}{2}\left(j_a^\mu + j_{5a}^\mu\right), \tag{5.8.6}$$

and in particular

$$u\gamma^\mu(1+\gamma_5)s \leftrightarrow 2(j_{L4}^\mu + ij_{L5}^\mu). \tag{5.8.7}$$

To evaluate this in terms of explicit pseudoscalar meson fields, we note that

$$\underset{\sim}{\Pi} = \frac{M}{\sqrt{2}}, \tag{5.8.8}$$

where M is the $SU(3)$ matrix, (4.8.7). We are only interested (for now) in the terms of (5.7.7) involving one K field and zero, and one or two pion fields. Here we will write down the terms involving the charged K field:

$$\begin{aligned}\overline{u}\gamma^\mu(1+\gamma_5)s &\to -\sqrt{2}\,f\partial^\mu K^- - i\left\{\tfrac{1}{\sqrt{2}}K^-\partial^\mu\pi^0 - \tfrac{1}{\sqrt{2}}\pi^0\partial^\mu K^-\right\} \\ &+ \tfrac{2}{3f}\left\{\tfrac{1}{\sqrt{2}}K^-\pi^+\partial^\mu\pi^0 - \sqrt{2}K^-\pi^-\partial^\mu\pi^+\right. \\ &- \left.\tfrac{1}{2\sqrt{2}}K^-\pi^0\partial^\mu\pi^0 + \tfrac{1}{2\sqrt{2}}\pi^0\pi^0\partial^\mu K^- + \tfrac{1}{\sqrt{2}}\pi^+\pi^-\partial^\mu K^-\right\} + \cdots\end{aligned} \tag{5.8.9}$$

To calculate the decay amplitudes, we now have only to calculate the matrix elements of (5.8.9) between the initial and final hadronic states in a theory with the interactions given by (5.4.9). To calculate to lowest order in an expansion in momentum, we need only include tree diagrams. The reason is that any loop diagram can be expanded in powers of momentum. The leading term is just a renormalization of the corresponding tree diagram. The other terms involve more powers of the momenta and are irrelevant in the approximation in which we are working. Thus, the effect of the loop diagram is included just by using renormalized parameters in the current and the Lagrangian (see Section 5.8).

In Feynman–diagram language, the amplitudes are shown in Figure 5–1, where the x represents the current (5.8.9).

Notice that the tree diagram contributions involve not only the current (5.8.9) but also the nontrivial interactions terms in the nonlinear Lagrangian (5.4.9). Both are required to make the matrix element of the current divergenceless in the limit $M = 0$ (see Problem 5–3).

We have seen that to lowest order in powers of momentum, the semileptonic K decays are completely determined by the chiral $SU(3) \times SU(3)$ symmetry. The predictions of Figure 5–1 are in fairly good agreement with the observed decay rates.

$K^- \to \mu^- \nu$

$K^- \to \pi^0 \mu^- \nu$

$K^- \to \pi^0 \pi^0 \mu^- \nu$

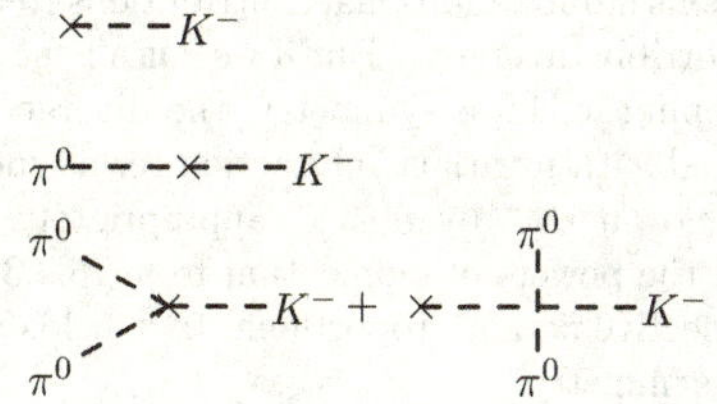

Figure 5-1:

5.9 The Chiral Symmetry–Breaking Scale

Why should it work? To understand the success of (5.4.9) as a description of the interactions of the pseudoscalar mesons, we must discuss the corrections to (5.4.9) in a systematic way. These arise from $SU(3) \times SU(3)$ invariant couplings involving more powers of derivatives or quark masses. The next most important terms, at low momentum, have one additional power of D^2 or μM compared to (5.4.9), such as the following:

$$\begin{gathered} \operatorname{tr}\left(D^\mu D^\nu U^\dagger D_\mu D_\nu U\right) \\ \operatorname{tr}\left(D^\nu U^\dagger D_\nu U U^\dagger \mu M\right) \\ \operatorname{tr}\left(U^\dagger \mu M U^\dagger \mu M\right) \end{gathered} \tag{5.9.1}$$

Terms with more powers of D^2 or μM are even less important.

We might expect that powers of D^2 and μM will be associated with inverse powers of some dimensional parameter that characterizes the convergence of the momentum expansion. We will call this parameter $\Lambda_{\rm CSB}$, the chiral symmetry–breaking scale. Thus, the terms (5.9.1) would have coupling constants of order $f^2/\Lambda_{\rm CSB}^2$, down by two powers of $\Lambda_{\rm CSB}$ compared to (5.4.9). In general, we would expect a term involving $2n$ derivatives and m powers of μM to have a coupling of order

$$f^2/\Lambda_{\rm CSB}^{2(n+m)}. \tag{5.9.2}$$

If these guesses are correct, then the convergence of the momentum expansion is determined by the size of the chiral symmetry–breaking scale. If $\Lambda_{\rm CSB}$ is large compared to the pseudoscalar meson masses, the higher order terms are negligible compared to (5.4.9) at any momentum involved in weak decays.

Unfortunately, it is unreasonable to suppose that $\Lambda_{\rm CSB}$ is very large because the radiative corrections to (5.4.9) produce terms like (5.9.1) and higher–order terms with infinite coefficients. Thus, even if these terms are absent for some choice of renormalization scale, they will be there for another. Under such circumstances, it would clearly be unreasonable to assume that the couplings (5.9.2) are small compared to the changes induced in them by a change in renormalization scale by some factor of order 1.

Let us explore these ideas in the example of π–π scattering. (5.9.10) predicts a four–pion vertex proportional to two powers of momentum (or one power of μm — we will concentrate on the momentum dependence). Schematically, it is

$$\left(\frac{p^2}{f^2}\right)\pi^4. \tag{5.9.3}$$

The quantum corrections induce additional contributions to π–π scattering from one–loop diagrams. The loop integral is horribly divergent. But if we cut off the divergent integral with some procedure that preserves the nonlinear chiral symmetry, the divergent contributions to the effective action should all be associated with invariant interaction terms such as (5.4.9) or (5.9.1).

There should be no quartic divergence, appropriately defined, because a quartic divergence could arise only if all the powers of momentum from (5.9.3) were loop momenta. But that would yield a term in the effective action proportional to π fields with no derivatives, which must vanish because of the chiral symmetry.

There is a quadratically divergent contribution proportional to two powers of external momenta. Schematically, this contribution is

$$\frac{1}{(4\pi)^2}\frac{p^2}{f^4}\Lambda_{\text{CO}}^2\pi^4, \tag{5.9.4}$$

where Λ_{CO} is a cut–off scale and the $1/(4\pi)^2$ is associated with the loop–momentum integration. This has the same form as (5.9.3), and so the quadratically divergent part can be absorbed into (5.9.3). It just renormalizes (5.4.9).

Finally, there is a logarithmically divergent term involving four powers of the external momenta. Schematically, it is

$$\frac{1}{(4\pi)^2}\frac{p^4}{f^4}\ln\left(\Lambda_{\text{CO}}/\kappa\right)\pi^4, \tag{5.9.5}$$

where κ is some arbitrary renormalization scale. This has the same form as the first term in (5.9.1). A change in the renormalization scale of order 1 (for example, e), will thus induce a change in (5.9.5) of order

$$\frac{1}{(4\pi)^2}\frac{p^4}{f^4}\pi^4, \tag{5.9.6}$$

which corresponds to a change in the coupling

$$f^2/\Lambda_{\text{CSB}}^2 \tag{5.9.7}$$

of order $1/(4\pi)^2$. Thus, we cannot reasonably assume that

$$f^2/\Lambda_{\text{CSB}}^2 << 1/(4\pi)^2, \tag{5.9.8}$$

because if this were true for one renormalization scale, it would not be true for another. But we could imagine that

$$f^2/\Lambda_{\text{CSB}}^2 \approx 1/(4\pi)^2. \tag{5.9.9}$$

If this were true in any reasonable renormalization scheme, it would be true in others. Indeed, (5.9.9) seems a reasonable guess for the chiral symmetry–breaking scale. If we assume

$$\Lambda_{\text{CSB}} \approx 4\pi f, \tag{5.9.10}$$

and cut off our loop integrals at momentum of order ΛCSB so that ΛCO $\approx \Lambda_{\text{CSB}}$, then the quantum corrections are of the same order of magnitude as the renormalized interaction terms. You can see this explicitly in (5.9.5) and (5.9.6) and can easily prove it in general. This is what we would expect on the basis of dimensional analysis in a strongly interacting theory in which there is no relevant small adjustable parameter. The factor of 4π in (5.9.10) comes from the dimensionality of space–time and is simply the appropriate connection between the constant f which appears in U

and the scale $\Lambda_{\rm CSB}$ which appears in loop effects. Hence, we will adopt (5.9.10) as a guess for the chiral symmetry–breaking scale.[1]

If these ideas make sense, we can understand the success of (5.4.9). $4\pi f$ is about 1 GeV, considerably larger than the typical momentum in K decay and more than a factor of two larger than the K mass. Thus we may hope that neglect of terms like (5.9.1) in our nonlinear Lagrangian will be a good approximation. Similarly, we can ignore all divergent loop effects.

Many of the ideas in this section were stimulated by a very nice paper by S. Weinberg (*Physica* **96A**:327–340 1979).

5.10 Important Loop Effects

Loop effects in effective chiral theories can often be ignored, as we have shown. But there are situations in which the loop effects are important because they can be qualitatively distinguished from the tree–diagram effects. All of these situations arise from the analytic structure of the loop effects. The most important are loops that produce infrared logs and imaginary parts.

Consider, as a simple example, π–π scattering amplitude in the chiral symmetry limit. The contribution from the diagram (for massless pions)

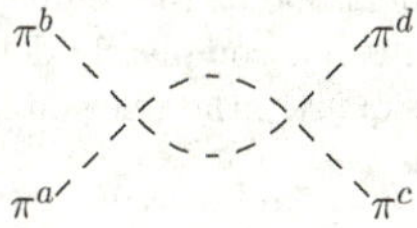

$$+\text{ cross terms} \tag{5.10.1}$$

$$\begin{aligned} = \Big\{ -\delta_{ab}\delta_{cd}\frac{s^2}{32\pi^2} - \delta_{ac}\delta_{bd}\frac{3s^2+u^2-t^2}{196\pi^2} \\ -\delta_{ad}\delta_{bc}\frac{3s^2+t^2-u^2}{196\pi^2}\Big\} \ln(-s/\kappa^2) \\ +\text{cross terms} + \text{polynomial in } s^2,\, t^2,\, u^2. \end{aligned} \tag{5.10.2}$$

In (5.10.2), s, t, and u are the Mandelstam variables, $s=(p_a+p_b)^2$, $t=(p_a-p_c)^2$, $u=(p_a-p_d)^2$; and κ is an arbitrary renormalizable scale. The polynomial comes from the matrix elements of interaction terms involving terms with four derivatives. Its coefficients cannot be calculated. But the logarithmic terms cannot arise from tree diagrams, so they are completely determined.

The real part of the $\ln(-s)$ term in (5.10.2) is an infrared logarithm. It diverges as the momenta go to zero. At very small momenta, it is more important than the terms with four derivatives from tree diagrams because the logarithm is large.

The imaginary part of the $\ln(-s)$ term in (5.10.2) arises from the unitarity of the S matrix. It is related to the total cross section.

Infrared logarithms can also play a role in the small symmetry breaking terms. For example, there is a contribution to the effective action with two derivatives (coming from a single Goldstone–boson loop) that has the form

$$\frac{f^2}{64\pi^2}\left(m^2\ln m^2/\kappa^2\right)_{ab}\cdot\frac{\delta^2}{\delta\pi_a\delta\pi_b}\mathrm{tr}\left(D_\mu U D^\mu U^\dagger\right), \tag{5.10.3}$$

[1]Note that there are also corrections to the lowest–order chiral Lagrangian from things like ρ exchange, which are not obviously related to chiral loop effects. But these are down by $1/m_\rho^2$, which is again $\simeq 1/\Lambda_{\rm CSB}^2$.

where M^2 is Goldstone–boson mass–squared matrix

$$m_{ab}^2 = \text{tr}\,(\{T_a,\, T_b\}\mu M)\,.$$

In principle, for light Goldstone bosons, this is the most important correction to the kinetic energy term (5.2.12) because the logarithm is large when κ is any reasonable renormalization scale ($\sim \Lambda_{\text{CSB}}$). In practice, however, they are not tremendously useful. If the logarithm is large, the symmetry breaking is small anyway. Thus, the terms proportional to $m_\pi^2 \ln m_\pi^2$ are very small while the terms proportional to $m_K^2 \ln m_K^2$ or $m_\eta^2 \ln m_\eta^2$ are no more important than the terms without logarithms, because $m_\kappa/\Lambda_{\text{CSB}}$ and $m_\eta/\Lambda_{\text{CSB}}$ are not much smaller than 1.

The general subject of renormalization in theories of this kind is a beautiful one which is part of the modern treatment, which is called, CHIRAL PERTURBATION THEORY. For more information, I suggest that you look in a very nice review of the subject (by the same name) by G. Ecker, Progress in Particle and Nuclear Physics, Vol. 35, 1995, Pergamon Press, Oxford.

5.11 Nonleptonic K Decay

The nonleptonic K decays, K^- into $\pi^0\pi^-$, $\pi^-\pi^-\pi^+$, and $\pi^0\pi^0\pi^-$, K_s into $\pi^+\pi^-$ and $\pi^0\pi^0$, and K_L into $3\pi^0$ and $\pi^+\pi^-\pi^0$, are not as well understood as the semileptonic and leptonic decays. The reason is that the form of decay Hamiltonian is not completely determined by the $SU(3) \times SU(3)$ symmetry. In tree approximation in terms of quarks, the relevant term is

$$\overline{u}\gamma^\mu(1+\gamma_5)s\overline{d}\gamma_\mu(1+\gamma_5)u. \tag{5.11.1}$$

Formally, this is just the product of two currents. But QCD corrections can change it in important ways, as we will see in detail later. From the point of view of the chiral symmetry, all we can say comes from the symmetry properties. (5.11.1) is a linear combination of terms that transform like an 8 and a 27 under the $SU(3)_L$. We can form several terms with the right transformation properties, and their renormalization is completely undetermined.

Nevertheless, we can say something. All of the terms we can build with right symmetry have a simple property that we can use to relate the $K^- \to 2\pi$ and 3π decays. We will use this system to exemplify another method for obtaining chiral symmetry predictions. The technique is called "current algebra".

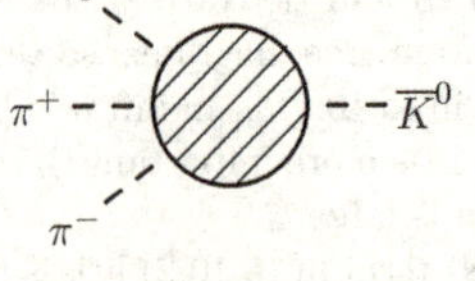

Figure 5-2:

Consider a $K \to 3\pi$ amplitude, for example

$$\langle \pi^+\pi^-\pi^0|H_W|\overline{K^0}\rangle = A(p_+,\, p_-,\, p_0), \tag{5.11.2}$$

where H_W is the term in the weak Hamiltonian responsible for the decay and the p's are the pion momenta, in an obvious notation. Now for reasons that are not entirely obvious, we consider the following:

$$\int e^{ip_0x}\,\langle \pi^+\pi^-|T\,(j_{53}^\mu(x)\,H_W)\,|K^0\rangle\, d^4x = A^\mu(p_+,\, p_-,\, p_0), \tag{5.11.3}$$

where j_{53}^{μ} is the axial isospin current. We will analyze this matrix element in the chiral symmetry limit, and calculate the quantity

$$\lim_{p_0\to 0} p_0^{\mu} A_{\mu}(p_+, p_-, p_0). \tag{5.11.4}$$

This can be calculated in two different ways. The p_0^{μ} can be turned into a derivative and taken inside the time–ordered product. The current is conserved, but we are left with an equal–time commutator

$$i\int e^{ip_0x}\,\delta(x^0)\,\langle\pi^+\pi^-|[j_{53}^0(x),\,H_W]|\overline{K^0}\rangle\,d^4x = -\frac{i}{2}\langle\pi^+\pi^-|H_W|\overline{K^0}\rangle. \tag{5.11.5}$$

The form of the equal–time commutator follows purely from the symmetry. The other way of evaluating (5.11.4) relies on the fact that the limit vanishes except for contributions from the pion pole due to the diagram shown in Fig 5.2. This contribution is

$$iF_{\pi}A(p_+, p_-, 0). \tag{5.11.6}$$

Thus, the 3π amplitude is related to the 2π amplitude. This relation works pretty well and is a good example of old–fashioned current algebra.

The effective Lagrangian technique is a systematic and reliable way of doing the same thing but, in simple cases like this, the current–algebra reduction formula can actually be faster.

Problems

5-1. If the axial $U(1)$ were a symmetry of QCD, there would be an $SU(3)$ singlet Goldstone-boson field, π_0, which in the $SU(3)$ symmetry limit transforms under a $U(1)$ transformation by a translation, $\delta\pi_0 = c_0$. In the presence of $SU(3)\times SU(3)$ symmetry breaking, the analog of (5.4.9) would be

$$\mathcal{L}(\pi) = \frac{1}{2}\partial^{\mu}\pi_0\partial_{\mu}\pi_0 + f^2\Big\{\frac{1}{4}\mathrm{tr}\,(\partial^{\mu}\Sigma^{\dagger}\partial_{\mu}\Sigma)$$
$$+\frac{1}{2}\mathrm{tr}\,\left(\mu M\Sigma^{\dagger}e^{-i\lambda\pi_0/f}\right) + \frac{1}{2}\mathrm{tr}\,\left(\mu M\Sigma e^{-i\lambda\pi_0/f}\right)\Big\} \tag{5-1}$$

where λ is an unknown constant that determines the strength of the spontaneous breaking of the $U(1)$ symmetry. Clearly when $M\to 0$, there are nine Goldstone bosons. Prove that (5-1) cannot describe the pseudoscalar mesons in our world for any value of λ. Assume $m_u = m_d = m$ and consider the 2×2 mass-squared matrix that describes π_0 and π_8. Show that the lightest isoscalar meson has a mass less than km_{π} for some k. What is k?

5-2. Prove (5.7.5).

5-3. Calculate the matrix element

$$\left\langle \pi^+\pi^- \,\middle|\, j_5^{\mu} \,\middle|\, K^-\right\rangle$$

where the K^-, π^+ and π^- momenta are P^{μ}, p_+^{μ}, and p_-^{μ}, respectively, and j_5^{μ} is the current corresponding to $\overline{u}\gamma^{\mu}\gamma_5 s$ in (5.8.9). Check your result by showing that the current is conserved in the chiral limit.

5-4. Suppose that (5.9.2) is satisfied for the coupling of higher order terms in the momentum expansion. Construct a power-counting argument to show that if a divergent loop diagram is cut off at momentum of order $\Lambda_{\text{CSB}} \approx 4\pi f$, it contributes terms to the effective action of the same order of magnitude as the bare interaction terms. Hint: Restrict yourself to the derivative interactions. Terms with derivatives replaced by μM clearly act the same way. Then a vertex has the form

$$(2\pi)^4\,\delta^4(\Sigma p_i)\,f^2\,p^2\left(\frac{p^2}{\Lambda^2_{\text{CSB}}}\right)^n\left(\frac{1}{f^2}\right)^n$$

where p is a typical momentum and m is the number of π lines emanating from the vertex. Explain this, then take N such vertices for various n and m. Call the loop momenta k and external momenta p. For each internal meson line, include a factor

$$\frac{1}{k^2}\,\frac{d^4k}{(2\pi)^4}$$

Suppose there are l such lines. Consider a contribution in which there is a k^{2j} from the momentum factors and the rest involves only external momenta, p^{2M}, for $M = N + \Sigma n - j$. Take it from there.

5-5. Write down the terms involving a $\overline{K^0}$ field and 0, 1, or 2 pions, in the current, (5.8.9).

5-6. Use chiral Lagrangian arguments to relate the following two matrix elements:

$$\left\langle \pi^+\pi^0 \,\middle|\, \bar{s}\gamma^\mu(1+\gamma_5)u\,\bar{u}\gamma_\mu(1+\gamma_5)d \,\middle|\, K^+ \right\rangle$$

and

$$\left\langle \overline{K^0} \,\middle|\, \bar{s}\gamma^\mu(1+\gamma_5)d\,\bar{s}\gamma_\mu(1+\gamma_5)d \,\middle|\, K^0 \right\rangle$$

Be sure to justify any assumptions you make about the form of the operators in the effective chiral theory.

6 — Chiral Lagrangians — Matter Fields

6.1 How Do the Baryons Transform?

Now that we have learned how to construct nonlinear Lagrangians that incorporate $SU(3) \times SU(3)$ symmetry for Goldstone bosons, we would like to do the same thing for the rest of the low-lying hadrons. I will illustrate some of the issues involved by considering the $SU(3)$ octet of spin-$\frac{1}{2}$ baryons.

To construct a theory describing the baryons, we must solve two problems. We need a sensible transformation law for the baryon fields, and we must somehow deal with the fact that masses of the baryons are not small compared to the chiral symmetry breaking scale. In this chapter, we will address the first of these issues and we will apply what we learn to a different problem — the constituent mass of the quarks.

We have some freedom to choose how the baryon fields transform under chiral $SU(3) \times SU(3)$. We might, for example, decide that the left- and right-handed baryon fields transform like an octet under $SU(3)_L$ and $SU(3)_R$, respectively. We could describe such baryons by chiral traceless 3×3 matrix fields Ψ_{1L} and Ψ_{1R} that transform as follows:

$$\Psi_{1L} \rightarrow L\Psi_{1L}L^\dagger, \quad \Psi_{1R} \rightarrow R\Psi_{1R}R^\dagger. \tag{6.1.1}$$

Alternatively, we could take both the left- and right-handed baryons to transform like octets under $SU(3)_L$. We could describe such baryons by a four–Dirac–component traceless 3×3 matrix field Ψ_2 that transforms as follows:

$$\Psi_2 \rightarrow L\Psi_2L^\dagger. \tag{6.1.2}$$

We could also take the baryons (left-handed or right-handed, or both) to transform under $SU(3) \times SU(3)$ like a $(3, \overline{3})$, described by a 3×3 matrix field Ψ_3 that transforms as follows:

$$\Psi_3 \rightarrow L\Psi_3R^\dagger. \tag{6.1.3}$$

There are eight such possibilities. We can take the left- and right-handed baryon fields independently to transform in any of four ways, like

$$(8,1),\ (1,8),\ (3,\overline{3}),\ \text{or}\ (\overline{3},3)$$

under chiral $SU(3) \times SU(3)$.

It might seem that each of these possibilities would lead to different physics. Not only do they transform differently under $SU(3) \times SU(3)$, but they even have different numbers of independent components. A $(3, \overline{3})$ field cannot be traceless, so it has nine components, while $(8, 1)$ or $(1, 8)$ has eight components.

Nevertheless, it doesn't matter which representation of baryons we choose. We can describe the low–momentum interactions of baryons and Goldstone bosons equally well with any of them. The reason is that the Goldstone boson matrix U can be used to transform from one representation to another. For example, we can take

$$\Psi_{2L} = \Psi_{1L}, \quad \Psi_{2R} = U\Psi_{1R}U^\dagger. \tag{6.1.4}$$

This gives the correct transformation properties for Ψ_2. For Ψ_3, although we cannot require that the trace vanish because $\operatorname{tr}\Psi_3$ is not invariant, we can require

$$\operatorname{tr}\left(U^\dagger\Psi_3\right) = 0. \tag{6.1.5}$$

Then we can take

$$\Psi_3 = \Psi_2 U, \tag{6.1.6}$$

which satisfies both (6.1.3) and (6.1.5).

The point is that the $SU(3) \times SU(3)$ symmetry is spontaneously broken down to $SU(3)$. In the low energy theory, it is only the $SU(3)$ transformation properties of the baryons that matter. Ψ_1, Ψ_2, and Ψ_3 (once (6.1.6) is imposed) all transform like $SU(3)$ octets.

6.2 A More Elegant Transformation Law

I argued in the previous section that we can build a nonlinear chiral theory with baryons in any $SU(3) \times SU(3)$ representation that reduces to an octet under the unbroken $SU(3)$ subgroup. Thus, we can decide on a baryon representation that is maximally convenient. All of the representations we have discussed so far are slightly inconvenient.

Ψ_2 and Ψ_3 are inconvenient because the representations are not parity invariant. Parity can be imposed, but the parity transformation is nonlinear because it involves the Goldstone–boson matrix U,

Ψ_1 is slightly inconvenient for a different reason. We saw in Chapter 5 that, in the limit of exact chiral symmetry, the Goldstone bosons have only derivative interactions because the chiral–invariant terms with no derivatives are just constants. This is an important feature of Goldstone–boson interactions. You can see it easily for the couplings of baryons to Goldstone bosons in the representation Ψ_2. To couple to baryons, the Goldstone bosons must be in an $SU(3)_R$ singlet because the baryons transform only under $SU(3)_L$. But if there are no derivatives, every $SU(3)_R$ invariant function of the Goldstone–boson fields is constant, because with a local $SU(3)_R$ transformation, all the π's in U can be transformed away. But in the parity invariant Ψ_1 representation, this feature of the Goldstone–boson interactions is not explicit. In this representation, the $SU(3) \times SU(3)$ invariant baryon mass term looks like

$$\operatorname{tr}\left(\overline{\Psi}_{1L} U \Psi_{1R} U^\dagger\right) + \text{h.c.} \tag{6.2.1}$$

In addition to the $SU(3)$ invariant baryon mass term, (6.2.1) describes some π–baryon couplings.

Of course, there is nothing really wrong with this representation. The pseudoscalar π–baryon coupling is equivalent to an axial–vector derivative coupling. But it might be nice to see it coming out explicitly. We can do this as follows. Define the matrix ξ as

$$\xi = e^{i\underset{\sim}{\Pi}/f}, \tag{6.2.2}$$

so that ξ is a square root of U,

$$\xi^2 = U. \tag{6.2.3}$$

The transformation properties of the $\underset{\sim}{\Pi}$ fields (5.2.1)–(5.2.4) induce a well-defined transformation of ξ under an $SU(3) \times SU(3)$ transformation. We can write it as

$$\xi \to \xi' = L\xi u(\pi)^\dagger = u(\pi)\xi R^\dagger . \tag{6.2.4}$$

This defines $u(\pi)$ as a nonlinear function of L, R, and $\underset{\sim}{\Pi}$.

The matrix $u(\pi)$ is invariant under the parity transformation $L \leftrightarrow R$, $\underset{\sim}{\Pi} \to -\underset{\sim}{\Pi}$. If $\underset{\sim}{c} = 0$ (that is, $L = R$), then the critical transformation is just an ordinary $SU(3)$ transformation, and $u(\pi)$ is the associated $SU(3)$ matrix ($u(\pi) = L = R$). If $c_a \neq 0$, then $u(\pi)$ depends on $\underset{\sim}{\epsilon}$, $\underset{\sim}{c}$, and $\underset{\sim}{\Pi}$ in a complicated nonlinear way.

This is the desired connection between the broken and unbroken subgroups. An arbitrary chiral transformation is represented by an ordinary $SU(3)$ matrix

$$u(\pi) = e^{i\underset{\sim}{v}} , \tag{6.2.5}$$

where the Hermitian matrix

$$\underset{\sim}{v}(\underset{\sim}{\epsilon},\ \underset{\sim}{c},\ \underset{\sim}{\Pi}) \tag{6.2.6}$$

encodes the chiral $SU(3) \times SU(3)$ in a nonlinear way.

Now we take our baryon fields Ψ to transform as

$$\Psi \to u(\pi)\Psi u(\pi)^\dagger . \tag{6.2.7}$$

The form of (6.2.7) is determined by the transformation properties of the baryon field under ordinary $SU(3)$. All of the chiral nature of the transformation is contained in $u(\pi)$. Note that $u(\pi)$ depends implicitly on the space–time coordinate x because it depends on $\underset{\sim}{\Pi}(x)$. In the transformation (6.2.7), the $u(\pi)$, Ψ, and $u(\pi)^\dagger$ fields are all to be evaluated at the same x. Thus, this chiral transformation is a local transformation. Derivatives of Ψ will not transform simply.

To see what is going on more explicitly, calculate $\underset{\sim}{v}$ for an infinitesimal chiral transformation, given by (5.2.5) for $\underset{\sim}{\epsilon}$ and $\underset{\sim}{c}$ infinitesimal. The result is (see Problem 6–1)

$$\underset{\sim}{v} = \underset{\sim}{\epsilon} + i[\underset{\sim}{c},\ \underset{\sim}{\Pi}]/f + 0(\pi^2). \tag{6.2.8}$$

For a chiral transformation $\underset{\sim}{c} \neq 0$, the transformation (6.2.7) multiplies the baryon field by functions of $\underset{\sim}{\Pi}$. This is an explicit realization of the vague statement that chiral transformations are associated with emission and absorption of Goldstone bosons.

It must be admitted that the advantages of this notation over another parity–invariant definition, such as Ψ_1, are rather slight. The most important difference is psychological. In this notation, all the Goldstone boson interactions in the chiral–symmetry limit are derivative interactions and obviously on the same footing. Whereas, if the baryons are described by Ψ_1, some are derivative interactions while some come from (6.2.1), and we may be misled into treating them differently (that is, assuming one set is small). At any rate, (6.2.2)–(6.2.7) define a simply and elegant notation, and we will use it.

Later, we will discuss the baryon Lagrangian that we can construct with this procedure. This involves an additional idea to deal with the large mass of the baryons (large, that is, compared to the chiral symmetry breaking scale) — the nonrelativistic effective theory — that we will introduce later. In the rest of this chapter, we will apply similar ideas to the quark model.

6.3 Nonlinear Chiral Quarks

In Section 3.2, I remarked that the nonrelativistic quark model gives a surprisingly useful description of the light baryon and meson states. In this section, I suggest a partial explanation of the success of the quark model. The problem is that the nonrelativistic quark model only makes sense if quarks have "constituent" masses, a few hundred MeV for the u and d quarks and about 500 Mev for the s quark. Whereas, we have learned from the arguments of Section 5.5 that the light quark masses that appear in the QCD Lagrangian are very different.

It seems reasonable to suppose that the bulk of the constituent mass of the light quarks is the effect of chiral symmetry breaking, like the baryon mass in the previous section. But then to incorporate $SU(3) \times SU(3)$, we must include Goldstone bosons and realize the chiral symmetry nonlinearly. Thus, "nonlinear chiral quarks" would seem to be a reasonable starting point in our search for an understanding of the success of the nonrelativistic quark model.

Let us, therefore, illustrate the ideas about thre transformation of matter field by building a matter Lagrangian describing the u, d, and s quark, in a triplet q, transforming as a triplet under the ordinary $SU(3)$. We will also assume that q is a triplet under color $SU(3)$, but for the moment we will not worry about the color $SU(3)$ gauge symmetry. Presumably, all derivatives should be color $SU(3)$ covariant derivatives, but since color $SU(3)$ and flavor $SU(3)$ commute, this shouldn't make much difference.

We require the quark field, q, to transform under $SU(3) \times SU(3)$ as

$$q \rightarrow u(\pi)q. \tag{6.3.1}$$

Then the term in a quark Lagrangian with no derivatives is unique:

$$\mathcal{L}_0^q = -m\overline{q}q. \tag{6.3.2}$$

The parameter m is the constituent mass of a (hypothetical) massless quark.

There are two ways to construct the chiral–invariant terms with a single derivative. We can use the ξ matrices to construct fields that transform linearly under $SU(3) \times SU(3)$:

$$\xi\, q_L \text{ and } \xi\, q_R = (3,1)\,, \qquad \xi^\dagger\, q_L \text{ and } \xi^\dagger\, q_R = (1,3)\,. \tag{6.3.3}$$

Note that because the chiral transformation properties come from the ξs and $\xi^\dagger$s, there is no correlation with the handedness of the quark fields. All of these fields transform only under global $SU(3) \times SU(3)$, so derivatives cause no special problem. We can simply put them together to form chiral invariant and parity invariant terms as follows:

$$\begin{aligned} &\tfrac{1+g_A}{2}\left[i\,\overline{q_L}\,\xi^\dagger \not\partial\,(\xi\, q_L) + i\,\overline{q_R}\,\xi \not\partial\left(\xi^\dagger\, q_R\right)\right] \\ &+\frac{1-g_A}{2}\left[i\,\overline{q_R}\,\xi^\dagger \not\partial\,(\xi\, q_R) + i\,\overline{q_L}\,\xi \not\partial\left(\xi^\dagger\, q_L\right)\right] \end{aligned} \tag{6.3.4}$$

An alternative method is to deal directly with the nonlinear local transformation (6.2.4)–(6.2.7). With the ξ matrices, we can build two vector fields with simple transformation properties. The vector field

$$\underset{\sim}{V}^{\,\mu} = \frac{1}{2}\left(\xi^\dagger\partial^\mu\xi + \xi\partial^\mu\xi^\dagger\right) \tag{6.3.5}$$

is a gauge field for the local transformation. It transforms as

$$\underset{\sim}{V}^{\,\mu} \rightarrow u(\pi)\underset{\sim}{V}^{\,\mu}u(\pi)^\dagger + u(\pi)\partial^\mu u(\pi)^\dagger. \tag{6.3.6}$$

The axial vector field

$$\underset{\sim}{A}^{\mu} = \frac{i}{2}\left(\xi^\dagger \partial^\mu \xi - \xi \partial^\mu \xi^\dagger\right) \tag{6.3.7}$$

is an $SU(3)$ octet field. It transforms as

$$\underset{\sim}{A}^{\mu} \to u(\pi) \underset{\sim}{A}^{\mu} u(\pi)^\dagger. \tag{6.3.8}$$

With the gauge field $\underset{\sim}{V}^{\mu}$, we can make a covariant derivative

$$D^\mu \Psi = \partial^\mu \Psi + \left[\underset{\sim}{V}^{\mu},\, \Psi\right], \tag{6.3.9}$$

which transforms homogeneously under the local transformation:

$$D^\mu \Psi \to u(\pi) D^\mu \Psi u(\pi)^\dagger. \tag{6.3.10}$$

With these tools, we can easily write down the terms with one derivative in the baryon Lagrangian

There are two terms involving one derivative:

$$\mathcal{L}_1^q = i\bar{q} \not{D} q + g_A \bar{q} \not{\underset{\sim}{A}} \gamma_5 q, \tag{6.3.11}$$

where

$$D^\mu = \partial^\mu + \underset{\sim}{V}^{\mu}, \tag{6.3.12}$$

and $\underset{\sim}{V}^{\mu}$ and $\underset{\sim}{A}^{\mu}$ are defined as follows:

If truncated at this point, the nonlinear chiral quark Lagrangian describes three degenerate quarks with a constituent mass and couplings to Goldstone bosons. The noteworthy thing about the Lagrangian is that is depends on only two new parameters m and g_A. As you will show in Problem 6–3, g_A is related to the renormalization of the quark contribution to the axial vector current. The point is that while the matrix elements of the vector $SU(3)$ currents are guaranteed to have their canonical form (at nonzero–momentum transfer) by the unbroken $SU(3)$ symmetry, there is no such constraint for the axial vector currents. $SU(3)$ considerations require them to have the form

$$g_A \bar{q} T_a \gamma^\mu \gamma_5 q + \cdots, \tag{6.3.13}$$

but the constant g_A is not determined by the symmetry. It is a parameter that must be calculated from the QCD dynamics (too hard!) or simply fit to the experiment. But this is the only freedom allowed in the nonlinear chiral quark theory to this order in the momentum expansion.

There are several terms that contain either two derivatives or one power of μM. We would expect, on the basis of the arguments (and assumptions) of Section 5.8 that these terms will have a "dimensional coupling" constant of the order of

$$m/\Lambda_{\text{CSB}}^2. \tag{6.3.14}$$

The most important such term is the symmetry–breaking term

$$-\frac{fm}{\Lambda_{\text{CSB}}^2} \bar{q}\left(\xi \mu M \xi + \xi^\dagger \mu M \xi^\dagger\right) q, \tag{6.3.15}$$

where f is a dimensionless constant or order 1. This term splits the s quark mass from the u and d masses by an amount

$$\Delta m \simeq \frac{2fmm_K^2}{\Lambda_{\rm CSB}^2} \simeq 150 \text{ MeV}. \tag{6.3.16}$$

The 150 MeV is a rough estimate of the mass difference from quark model phenomenology, for example, the mass splitting between states of different strangeness in the spin–$\frac{1}{2}$ baryon decuplet. Note that if the constituent mass m is about 350 MeV, then f is indeed of order 1.

In general, we should also consider interaction terms involving more than two quark fields. However, these are higher dimension operators than those that we have considered so far, and we might expect their coefficients to be suppressed by additional powers of $1/\Lambda_{\rm CSB}$. If we generalize the analysis of Section 5.8 to include quark couplings, we conclude that we should include an extra factor of $\Lambda_{\rm CSB}$ for each pair of quark fields. In practice, this makes the terms with additional quark fields less important. Their matrix elements are proportional (roughly) to powers of f_π, while their coefficients are suppressed by powers of $1/\Lambda_{\rm CSB}$. Thus, we will usually ignore such operators.

6.4 Successes of the Nonrelativistic Quark Model

The hope is that now that we have included the effect of chiral symmetry breaking in the constituent quark mass, the interquark forces will be correspondingly weakened in our effective theory. Thus, we may try to identify the low–lying hadrons with nonrelativistic bound states of the quarks described in Section 6.4. Presumably, the quarks are bound by confining QCD interactions, along with effects of multiquark and multigluon operators that appear in high orders of $1/\Lambda_{\rm CSB}$ in the effective Lagrangian.

This picture of the baryons is attractive for several reasons. The first striking success is that the baryon masses are given correctly by this picture. The spin–$\frac{1}{2}$ octet and spin–$\frac{3}{2}$ decuplet are interpreted as color–singlet bound states of three quarks in a ground–state wave function with total angular momentum equal to zero. The leading contribution to the baryon mass in the nonrelativistic limit is just the sum of the constituent quark masses. The most interesting relativistic correction is a spin–dependent contribution to the mass of the form of a sum of quark pairs

$$\kappa \sum_{ij} \frac{\vec{S}_i \cdot \vec{S}_j}{m_i m_j}, \tag{6.4.1}$$

where $\vec{S}_i$ and m_i are the spin and mass of the ith quark and κ is a parameter that depends on the detailed dynamics. (6.4.1) arises primarily from gluon exchange, which gives a contribution of the right sign. It splits the decuplet from the octet by a few hundred MeV. Furthermore, it explains the sign and magnitude of the $\Sigma - \Lambda$ difference.

A second success of the quark picture is the prediction of the magnetic moments. In leading order in v/c, the baryon moment is simply the sum of the contributions from the individual quarks,

$$\mu\vec{S} = \sum_i \mu_i \vec{S}_i = \sum_i Q_i \vec{S}_i / m_i, \tag{6.4.2}$$

where Q_i is the quark change. A good picture of the baryon masses is obtained if we take

$$m_u \simeq m_d \approx m = 360\,\text{MeV}, \quad m_s \simeq 540\,\text{MeV}, \tag{6.4.3}$$

(see Problem 6–5). With these masses, the octet baryon magnetic moments are those given in the second column in Table 6.1 (in nuclear magnetons). The agreement with the data shown in the third column is excellent.

Baryon	$\mu_{\text{Quark model}}$	$\mu_{\text{Exp}}{}^*$
p	2.61	2.79
n	-1.74	-1.91
Λ	-0.58	-0.61
Σ^+	2.51	2.36 ± 0.01
Σ^-	-0.97	-1.15 ± 0.02
Ξ^0	-1.35	-1.25 ± 0.02
Ξ^-	-0.48	-0.69 ± 0.04

Table 6.2: Baryon magnetic moments. * Source: L. C. Pondrom, *AIP Conference Proceedings*, No. 95, New York: AIP, 1982. See also articles by R. Handler, J. Mariner, and B. L. Roberts in the same publication.

The success of (6.4.2) in giving not only the ratios of the baryon magnetic moments, but even their overall scale, seems to me to be very significant. The interesting thing about it is that (6.4.2) depends on the approximate renormalizability of the effective chiral theory. Nonrenormalizable terms, such as terms describing anomalous magnetic moments for the quarks, complicate the expression for μ of the baryons. Thus, the success of (6.4.2) shown in Table 6.1 is an indication that these renormalizable terms are small. This is what we expect if Λ_{CSB} is of the order of 1 GeV. We would expect an anomalous magnetic moment term of the form

$$\frac{mf}{\Lambda_{\text{CSB}}^2} F_{\mu\nu}\bar{q}\sigma^{\mu\nu}q, \tag{6.4.4}$$

where f is of order 1. This gives an anomalous moment of order

$$m^2/\Lambda_{\text{CSB}}^2, \tag{6.4.5}$$

which we expect to be about 10%. Other nonrelativistic terms and relativistic corrections give similar contributions to the moments. Thus, in context, we can understand the success of the magnetic moment predictions of the quark model.

We can also use the quark model to calculate the matrix elements of the vector and axial currents. The f_1 values for the vector currents are guaranteed to work at zero momentum to the extent that the wave functions are $SU(3)$ symmetrical. The axial vector currents are more subtle. If we had built our quark model naively, without thinking about spontaneously broken chiral symmetry, we might have expected the axial currents to be given by

$$\bar{q}T_a\gamma^\mu\gamma_5 q. \tag{6.4.6}$$

This would give a value 5/3 for g_1 neutron decay, whereas the experimental value is 1.25, over 30% off. Thus, it is a good sign that in a properly constructed chiral theory, the normalization of the axial current can be different. The neutron decay value determines g_A to be

$$g_A = .75 \tag{6.4.7}$$

In the language of Section 6.3, the overall scale of D and F are related to g_A. But the ratio D/F must now be fixed because we have no further freedom to change the form of the axial vector current (see Problem 6–6). Again, this prediction works rather well.

We have seen that a chiral nonlinear quark model gives a very attractive picture of the properties of the octet and decuplet baryons. But this model is probably not the most useful way of thinking about the low–lying mesons. I say this because in a nonchiral quark model, one naturally interprets the π and the ρ, for example, as the spin–1 and spin–0 states of a quark–antiquark pair in a zero orbital angular–momentum state. As in the baryon sector, the spin–dependent interaction (6.4.1) splits these states. And it has the right sign to make the ρ heavier. If we bravely apply the same sort of nonrelativistic reasoning to this system, we get a nice simple picture of the mesons as well as the baryons.

In our picture, however, the spin–0 bound state of quark and antiquark is not the pion. The pion is an elementary Goldstone boson. Presumably, what happens is that s–channel pion exchange in the quark–antiquark interaction produces a repulsive force that pushes up the $q\overline{q}$ state. Thus, the connection between ρ and π is not very useful. The mystery of the connection between QCD and the quark model remains, at least for the meson states.

6.5 Hyperon Nonleptonic Decays

As an example of the application of chiral–quark–model ideas to weak interactions, consider the nonleptonic decays of the strange baryons, the hyperons Λ, Σ, and Xi. The observed decays are (along with the conventional shorthand):

$$\begin{aligned}
\Lambda^0_0 &: \Lambda \rightarrow \pi^0 n \\
\Lambda^0_- &: \Lambda \rightarrow \pi^- p \\
\Sigma^+_+ &: \Sigma^+ \rightarrow \pi^+ n \\
\Sigma^+_0 &: \Sigma^+ \rightarrow \pi^0 p \\
\Sigma^-_- &: \Sigma^- \rightarrow \pi^- n \\
\Xi^0_0 &: \Xi^0 \rightarrow \pi^0 \Lambda \\
\Xi^-_- &: \Xi^- \rightarrow \pi^- \Lambda .
\end{aligned} \tag{6.5.1}$$

We can write the invariant amplitude for a nonleptonic decay as

$$M = G_F m_\pi^2 [\overline{u}_f(p_f)(A + B\gamma_5)u_i(p_i)], \tag{6.5.2}$$

where $u_i(u_f)$ is the spinor describing the initial (final) baryon state with momentum $p_i(p_f)$ and m_π is the $\pi^\pm$ mass. The A amplitude describes the parity–violating s–wave decay, the B amplitude the parity–conserving p–wave.

All these decays are produced by the $\Delta S = 1$ weak Hamiltonian that, ignoring QCD and other radiative corrections, looks like

$$\frac{G_F}{\sqrt{2}} s_1 c_1 c_3 \overline{u}\gamma^\mu(1+\gamma_5)s\overline{d}\gamma_\mu(1+\gamma_5)u. \tag{6.5.3}$$

We will see in Chapter 9 how this is modified by QCD. At the moment, what is important is its transformation property under $SU(3)_L \times SU(3)_R$. It transforms as a linear combination of $SU(3)_L$ 8 and 27. In the effective chiral–quark Lagrangian, (6.5.3) will appear as a sum of operators that transform the same way. As with the terms in the Lagrangian, we can order the terms in increasing numbers of derivatives, symmetry–breaking parameters, and quark fields.

Before we discuss the chiral–quark model, it may be useful to set the historical context by discussing the nonleptonic decays in the chiral mode of baryons discussed in Sections 6.1–3. We will concentrate on the $SU(3)$ octet operator because, as we will see in Chapter 9, there is reason to believe that it is enhanced by the QCD interactions.

It is helpful to consider a matrix h,

$$h^i_j = \delta^i_2 \delta^3_j, \tag{6.5.4}$$

which is a symmetry–breaking parameter whose coefficient is the operator of interest. That is, is we imagine that h transforms under $SU(3)_L$ as

$$h \to L h L^\dagger, \tag{6.5.5}$$

and build all the $SU(3)_L \times SU(3)_R$ invariants that are linear in h, we will have the operators that we want. For example, in the chiral Goldstone–boson theory, the leading octet contributions are

$$\mathrm{tr}\left[h(\partial^\mu U)\partial_\mu U^\dagger\right] \quad \mathrm{tr}\left[h U \mu M\right], \tag{6.5.6}$$

which involve derivatives or explicit symmetry breaking (note that the second term can be absorbed into a redefinition of the π masses, so it does not contribute to decays).

The interesting thing about the chiral baryon theory is that there are operators that do not involve any derivatives or μM. There are two such operators involving only two baryon fields whose matrix elements are not proportional to momenta or μM. They can be written as

$$G_F m_\pi^2 f_\pi \mathrm{tr}\,\overline{\Psi}\left[d\{\xi^\dagger h \xi,\ \Psi\} + f[\xi^\dagger h \xi,\ \Psi]\right]. \tag{6.5.7}$$

There are other terms that involve no explicit derivatives, such as

$$\mathrm{tr}\,\overline{\Psi}\gamma_5 \xi^\dagger h \xi \Psi. \tag{6.5.8}$$

But their matrix elements vanish in the symmetry limit because of the γ_5. They contain an implicit factor of the symmetry breaking, or pion momentum.

It is reasonable to suppose that (6.5.7) gives the dominant contribution to the baryon nonleptonic decays (assuming octet enhancement), with the effect of operators like (6.5.8) suppressed by powers of $p_\pi/\Lambda_{\mathrm{CSB}}$. This assumption gives the standard current algebra results for the nonleptonic decays. It works very well for the s–wave amplitudes. For example, with

$$d = -\frac{1}{2}, \quad f = \frac{3}{2}, \tag{6.5.9}$$

the result for the s–wave decay amplitudes is shown in Table 6.2, along with the experimental value (where the signs have been chosen consistent with the conventions used in the particle data book). No fit has been attempted here, but it is clear that everything works to 10% or so, which is all we can expect from the effective theory (see Problem 6–7).

The theoretical predictions exhibit five relations among the seven amplitudes that follow from (6.5.7). Three of these are $\Delta I = \frac{1}{2}$ relations.

$$\begin{aligned} A(\Lambda^0_-) &= -\sqrt{2}\,A(\Lambda^0_0) \\ A(\Xi^-_-) &= -\sqrt{2}\,A(\Xi^0_0) \\ \sqrt{2}\,A(\Sigma^+_0) + A(\Sigma^+_+) &= A(\Sigma^-_-). \end{aligned} \tag{6.5.10}$$

Decay	A(Th)	A(exp)
Λ^0_-	1.63	$1.47 \pm .01$
Λ^0_0	-1.15	$-1.07 \pm .01$
Σ^+_+	0	$0.06 \pm .01$
Σ^+_0	1.41	$1.48 \pm .05$
Σ^-_-	2.00	$1.93 \pm .01$
Ξ^0_0	1.44	$1.55 \pm .03$
Ξ^-_-	2.04	$2.04 \pm .01$

Table 6.3: Baryon decay amplitudes.

One is the Lee-Sugawara relation

$$\sqrt{3}\,A(\Sigma^+_0) + A(\Lambda^0_-) = 2A(\Xi^-_-) \tag{6.5.11}$$

and one is the simple relation

$$A(\Sigma^+_+) = 0. \tag{6.5.12}$$

All are satisfied as well as we could expect (or better).

Unfortunately, we can now take (6.5.9) from the s–waves and predict the p–waves. The p–wave decays, in this picture, arise from so-called pole diagrams in which the hyperon changes strangeness through the $\overline{\psi}\psi$ term in (6.5.7) before or after a pion is emitted. The pion emission is accompanied by a factor of pion momentum, proportional to symmetry breaking. But this is compensated by the pole in the baryon propagator that is inversely proportional to the symmetry breaking, so the contribution is large. The only trouble is that the predictions for the p–wave decay look absolutely nothing like the data (see Problem 6–7). This has been a puzzle for nearly twenty years. Is this a general problem for the effective Lagrangian idea, or is there something more specific wrong with the assumption that (6.5.7) dominates the nonleptonic decays?

Can we find our way out of this difficulty in the chiral–quark model? As in the baryon theory, there are octet operators in the quark theory that do not involve derivatives or μM. With two quark fields there is a unique operator,

$$\overline{\psi}\xi^\dagger h\xi\psi. \tag{6.5.13}$$

With four quark fields, there are a variety of operators, such as

$$\overline{\psi}\xi^\dagger h\xi\gamma^\mu T_a\psi\overline{\psi}\gamma_\mu T_a\psi. \tag{6.5.14}$$

We expect (6.5.13) to be more important than (6.5.14), but because (6.5.13) is chirality–violating, the detailed power–counting analysis of Section 5.8 (and Problem 5–4) suggests that the suppression of (6.5.14) compared to (6.5.13) is only

$$\sim \frac{F_\pi}{m} \simeq \frac{1}{3}. \tag{6.5.15}$$

The most penetrating way to analyze the chiral–quark theory is to use it to construct a chiral–baryon theory by comparing the matrix elements of the corresponding interactions between various states. For example, the matrix elements that give rise to the s–wave amplitudes in the quark theory from the two–quark operator (6.5.13) are obtained by acting with $\overline{\psi}[h,\,\underset{\sim}{\Pi}]\psi$ on each of the quark lines. Any s quark can emit a π^0 or π^- and turn into a d or u. This produces a baryon matrix element like (6.5.7), but because the quark–flavor dependence is like that of an $SU(3)$ generator, the matrix is pure f. Thus (6.5.13) makes a contribution to (6.5.7) with

$$d = 0. \tag{6.5.16}$$

The four–quark operators like (6.5.14) produce similar terms, but with both f and d contributions. Thus, we expect a slight suppression of the d amplitude, by a factor of about $\frac{1}{3}$ (from (6.5.15)), which is just what is observed (in (6.5.9)).

The same terms, (6.5.7), (6.5.13), and (6.5.14) give rise to p–wave amplitudes through pole diagrams in both the baryon and quark theories. But here the correspondence is not exact. If we look at the baryon pole diagram in the quark–model language with the operator given by (6.5.14), the transition induced by (6.5.14) and the pion emission are uncorrelated events. For example, in the decay $\Lambda \to \pi^0 N$, in the pole diagram with an intermediate N, the π^0 emission amplitude is a coherent sum of π^0 emission from each of the three quarks in the N. But in the quark model, while all the same diagrams exist, the s to d transition and pion emission are not independent. Now we must keep track of the energy and momentum of each quark. Typically, the QCD binding forces must redistribute the energy produced in the transition in order for the decay to take place. For example, it clearly makes a difference whether the transition and the emission occur on the same quark line or not. The momentum redistribution that must take place to get from the initial–state to final–state baryons is quite different in the two cases. Thus, while the quark theory with operators (6.5.13) and (6.5.14) gives s–wave amplitudes that can be interpreted in a baryon theory as coming from the operator (6.5.7), there is no reason to believe that the p–wave amplitudes produced by (6.5.14) are the same as those produced by (6.5.7).

If this is correct, then one theory or the other, or both, must contain additional contributions to the $\Delta S = 1$ octet operator that makes up the difference. The leading contributions must involve an explicit Goldstone–boson field with vector coupling because the mismatch does not occur in the s–wave amplitudes. For example, in the baryon theory, there are the six such $SU(3)$ singlets that we can make by taking traces in various orders of the four octets

$$\overline{\Psi}, \quad \gamma_\mu\gamma_5\overline{\Psi}, \quad \xi^\dagger h \xi, \quad \underset{\sim}{A}^{\,\mu}, \tag{6.5.17}$$

where $\underset{\sim}{A}^{\,\mu}$ (defined by (6.3.7) is the chiral version of the derivative of the $\underset{\sim}{\Pi}$ field. Similar operators are possible in the quark theory.

These extra contributions are peculiar in the sense that they involve a derivative but are designed to produce an effect that is not proportional to the symmetry breaking. Their coefficients must be inversely proportional to the symmetry breaking. The moral is that we cannot avoid such terms in both theories simultaneously. They are certainly absent in the fundamental QCD–quark theory, and it is hard to see how they could get induced in the transition to the chiral–quark model. If (6.5.13) and (6.5.14) dominate in the chiral–quark model, then the extra operators with large coefficients will be induced in the baryon theory.

Of course, this doesn't make much sense in the limit in which the symmetry breaking goes to zero. When the quark mass differences become very small compared to typical momenta in the baryon bound state, the distinction between different energy and momentum redistributions in the decay disappears. Then the coefficients of the extra terms level off so that they can approach finite but large limits as the symmetry breaking vanishes.

The result of all this is what we expect, in the baryon theory, the operators constructed out of (6.5.17) to appear with large coefficients. Four of these contribute independently to the p–wave decays. This eliminates all relations for the p–wave amplitudes except the isospin relations, which are well satisfied.

In a sufficiently explicit quark model, it should be possible to find relations among some of these extra parameters and test these ideas in detail. For now, we will content ourselves with the negative result that the p–wave amplitudes cannot be simply predicted.

Problems

6-1. Derive (6.3.6).

6-2. With $\mathcal{L}^B = \mathcal{L}_0^B + \mathcal{L}_1^B$, calsulate the axial vector currents, and calculate their matrix elements between baryon states. Derive the Goldberger-Treiman relations. Use the meson Lagrangian including the chiral symmetry-breaking term.

6-3. Find the quark-current contribution to the vector and axial-vector currents from $\mathcal{L}^q = \mathcal{L}_0^q + \mathcal{L}_1^q$, (6.3.2) and (6.3.11). Derive a Goldberger-Treiman relation for constituent quark masses.

6-4. Find all the $SU(3) \times SU(3)$ invariant terms in the chiral-quark Lagrangian with two derivatives or one μM, and no more than two quark fields.

6-5. If we include only the spin-dependent relativistic corrections, the masses of the ground-state baryons have the form

$$\sum_i m_i + \kappa \sum_{i,j} \frac{\vec{S}_i \cdot \vec{S}_j}{m_i m_j}$$

Show that (6.4.3) gives a good fit to the octet and decuplet masses.

6-6. Use the chiral-quark model to find d/f in the chiral-baryon Lagrangian (6.5.7). Note that in the nonrelativistic limit, the matrix elements of j^μ_{5a} are nonzero only for the space components, and these have the form

$$\sum \text{quark}\, T_a\, \vec{\sigma}$$

6-7. Calculate the s-wave and p-wave amplitudes for the observed hyperon nonleptonic decays from (6.5.7) and compare with the data in the "Review of Particle Properties." You will need to understand their sign conventions. Find the four operators built from (6.5.17) that contribute to the observed decays. Calculate their contributions and find the coefficients such that together with (6.5.7) and (6.5.9) they give a reasonable picture of both s-wave and p-wave amplitudes.

6a - Anomalies

6a.1 Electromagnetic Interactions and $\pi^0 \to 2\gamma$

The π^0 decay into two photons obviously involves electromagnetic interactions, not weak interactions. Nevertheless, I cannot resist discussing it here. It affords a beautiful illustration of the chiral Lagrangian technique together with a good excuse to study some of the history of the axial anomaly (which we used in Section 5.4) to argue away the axial $U(1)$ symmetry).

It would seem fairly trivial to incorporate the effects of electromagnetism into our chiral Lagrangian. In the underlying QCD theory, we merely replace the usual color $SU(3)$ covariant derivative (3.1.3) with one that incorporates the photon field

$$D^\mu \to D^\mu + ieQA^\mu, \tag{6a.1.1}$$

where Q is the quark charge matrix

$$Q = \begin{pmatrix} \frac{2}{3} & 0 & 0 \\ 0 & -\frac{1}{3} & 0 \\ 0 & 0 & -\frac{1}{3} \end{pmatrix}. \tag{6a.1.2}$$

In our chiral Lagrangian, we can simply impose electromagnetic gauge invariance in the same way. For example, we replace $\partial^\mu U$ with

$$D^\mu U = \partial^\mu U + ieA^\mu \Big[Q, U\Big]. \tag{6a.1.3}$$

The commutator in (6a.1.3) arises because Q acts on both left-handed and right-handed quarks. That is, the electromagnetic charge is in the ordinary $SU(3)$ subgroup of $SU(3) \times SU(3)$.

The replacement (6a.1.3) has a variety of effects. In Section 5.5, for example, we discussed the effect of such interactions on the pseudoscalar masses. These arose because the electromagnetic interactions explicitly break the chiral (and ordinary) isospin and V–spin symmetries. We thus expect a variety of symmetry–breaking terms proportional to the quark charges. However, these classical electromagnetic effects are not very efficient in producing the decay $\pi^0 \to 2\gamma$.

From angular momentum and parity considerations, it is clear that the two photons in the π^0 decay are in an $l = 1$, $j = 0$ state associated with the pseudoscalar operator

$$\epsilon_{\mu\nu\lambda\sigma} F^{\mu\nu} F^{\lambda\sigma}. \tag{6a.1.4}$$

For example, an interaction of the form

$$\pi_3 \epsilon_{\mu\nu\lambda\sigma} F^{\mu\nu} F^{\lambda\sigma} \tag{6a.1.5}$$

would produce π^0 decay.

The pseudoscalar photon operator is contained in terms like

$$i\epsilon_{\mu\nu\lambda\sigma}\,\mathrm{tr}\,U^\dagger D^\mu D^\nu D^\lambda D^\sigma U + \mathrm{h.c.} \tag{6a.1.6}$$

However, such terms never contain the interaction (6a.1.5). The reason is the commutator (6a.1.3). The quark charge matrix QCD commutes with T_3 because both are diagonal. Quite generally, in the limit of exact chiral symmetry, terms like (6a.1.5) are impossible. This follows simply from the fact that, at least classically, electromagnetism doesn't break the chiral symmetry associated with the π_3 field. The argument should by now be familiar. If there is no derivative acting on the π fields, the unbroken chiral symmetries can be taken to be local, and the corresponding Goldstone bosons can be transformed away. Thus, as before, the Goldstone bosons have only derivative interactions. (6a.1.5) is ruled out!

This reasoning is called the Sutherland theorem, after David Sutherland who derived an equivalent statement using current algebra. It worried people for a long time because it seemed to imply that electromagnetism could not account for the $\pi^0 \to 2\gamma$ decay, at least if current–algebra reasoning was correct.

In our modern language, we can see that there are possible interaction terms with derivatives, such as

$$i\epsilon_{\mu\nu\lambda\sigma}\, F^{\mu\nu}\, F^{\lambda\sigma}\,\mathrm{tr}\left(Q^2 U^\dagger \partial^\alpha \partial_\alpha U\right) + \mathrm{h.c.} \tag{6a.1.7}$$

There are also terms involving the explicit chiral symmetry breaking, such as

$$i\epsilon_{\mu\nu\lambda\sigma}\, F^{\mu\nu}\, F^{\lambda\sigma}\,\mathrm{tr}\left(Q^2 \mu M U\right) + \mathrm{h.c.} \tag{6a.1.8}$$

But the contribution of these terms to the π^0 decay amplitude is suppressed by a power of m_π^2 (over Λ^2_{CSB}, presumably) compared to the contribution of a term like (6a.1.5). For any reasonable coefficient, they are just too small to account for the observed decay rate.

The resolution of this puzzle is the anomaly. In fact, the chiral symmetry associated with π_3 is broken by the electromagnetic interaction, but only through a triangle diagram like (5.5.4) with the gluons replaced by photons. The axial T_3 current is *not* conserved. Rather it satisfies

$$\partial_\mu j_5^\mu = \frac{3e^2}{16\pi^2}\mathrm{tr}\left(T_3 Q^2\right)\cdot \epsilon_{\mu\nu\lambda\sigma}\, F^{\mu\nu}\, F^{\lambda\sigma}. \tag{6a.1.9}$$

The factor of 3 in (6a.1.9) comes from the sum over the three colors of the quarks. Thus, there must be an extra term in our chiral Lagrangian with the property that under a T_3 chiral transformation this extra term changes by

$$-\,c_3\frac{3e^2}{16\pi^2}\mathrm{tr}\left(T_3 Q^2\right)\epsilon_{\mu\nu\lambda\sigma}\, F^{\mu\nu}\, F^{\lambda\sigma}. \tag{6a.1.10}$$

The extra term must also have appropriate transformation properties under all the other $SU(3)_L \times SU(3)_R$ transformations. This is not trivial to implement. Wess and Zumino (*Physics Letters* **37B**:95–97, 1971) determined the form of the extra term by imagining a theory in which not only the electromagnetic $U(1)$ is gauged, but also the full $SU(3)_L \times SU(3)_R$, as we did in (5.3.1). The form of the anomaly in this theory, with all of its dependence on the non-Abelian structure of the group, is determined by the algebra of the infinitesimal gauge transformations. The anomaly equation can then be formally integrated to give the extra term.

Let

$$l^\mu = l_a^\mu T_a, \quad r^\mu = r_a^\mu T_a \tag{6a.1.11}$$

be conventionally normalized $SU(3)_L$ and $SU(3)_R$ gauge fields. Then an $SU(3)_L \times SU(3)_R$ gauge transformation in the effective theory has the form:

$$\begin{aligned} &U \to LUR^\dagger \\ &l^\mu \to Ll^\mu L^\dagger - iL\partial^\mu L^\dagger \\ &r^\mu \to Rr^\mu R^\dagger - iR\partial^\mu R^\dagger \end{aligned} \tag{6a.1.12}$$

It is convenient to work with the vector and axial vector fields that we introduced in (5.3.2),

$$v^\mu = \left(l^\mu + r^\mu\right)/2, \quad a^\mu = \left(l^\mu - r^\mu\right)/2, \tag{6a.1.13}$$

which are gauge fields for the ordinary and chiral $SU(3)$ transformations, respectively. In the notation of Wess and Zumino, an infinitesimal $SU(3)$ gauge transformation looks as follows (with all space–time variables evaluated at the same point):

$$\begin{aligned} &L = R = 1 + i\underset{\sim}{\epsilon} \\ &\delta_\epsilon U = i\left[\underset{\sim}{\epsilon},\, U\right] \\ &\delta_\epsilon v^\mu = i\left[\underset{\sim}{\epsilon},\, v^\mu\right] - \partial^\mu \underset{\sim}{\epsilon} \\ &\delta_\epsilon a^\mu = i\left[\underset{\sim}{\epsilon},\, a^\mu\right] \end{aligned} \quad . \tag{6a.1.14}$$

An infinitesimal chiral gauge transformation looks as follows:

$$\begin{aligned} &L = R^\dagger = 1 + i\underset{\sim}{c} \\ &\delta_c U = i\left[\underset{\sim}{c},\, U\right] \\ &\delta_c v^\mu = i\left[\underset{\sim}{c},\, a^\mu\right] \\ &\delta_c a^\mu = i\left[\underset{\sim}{c},\, v^\mu\right] - \partial^\mu \underset{\sim}{c} \end{aligned} \quad . \tag{6a.1.15}$$

In this notation, the commutation relations of the $SU(3)_L \times SU(3)_R$ gauge algebra take the form

$$\begin{aligned} \delta_{\epsilon_1}\delta_{\epsilon_2} - \delta_{\epsilon_2}\delta_{\epsilon_1} &= \delta_{\epsilon_3}, & \underset{\sim}{\epsilon}_3 &= -i\left[\underset{\sim}{\epsilon}_1,\, \underset{\sim}{\epsilon}_2\right] \\ \delta_\epsilon\delta_{c_1} - \delta_{c_1}\delta_\epsilon &= \delta_{c_2}, & \underset{\sim}{c}_2 &= -i\left[\underset{\sim}{\epsilon},\, \underset{\sim}{c}_1\right] \\ \delta_{c_1}\delta_{c_2} - \delta_{c_2}\delta_{c_1} &= \delta_\epsilon, & \underset{\sim}{\epsilon} &= -i\left[\underset{\sim}{c}_1,\, \underset{\sim}{c}_2\right]. \end{aligned} \tag{6a.1.16}$$

In the absence of the anomaly, the algebra (6a.1.16) is realized trivially on the effective action of the chiral theory, which is simply invariant. But the extra term that we must add to incorporate the effect of the anomaly is not annihilated by the δ's. Call this term $W(U,\, v^\mu,\, a^\mu)$. If we renormalize the theory so as to maintain invariance under ordinary $SU(3)$ gauge transformations, then

$$\delta_\epsilon W = 0, \quad \delta_c W = F\Big(\underset{\sim}{c},\, v^\mu,\, \underset{\sim}{a}^\mu\Big). \tag{6a.1.17}$$

Now, (6a.1.17), (6a.1.16), and the Abelian anomaly (6a.1.10) are sufficient to determine the full non-Abelian anomaly. The result of a straightforward but tedious calculation is (again *ala* Wess and Zumino)

$$F\Big(\underset{\sim}{c},\, v^\mu,\, a^\mu\Big) = \int f\Big(\underset{\sim}{c}(x),\, v^\mu(x),\, a^\mu(x)\Big)\, d^4x\,, \tag{6a.1.18}$$

where

$$\begin{aligned} f = -\frac{1}{16\pi^2}\epsilon_{\mu\nu\alpha\beta}\,\mathrm{tr}\,\Big(\,&\underset{\sim}{c}\Big(3v^{\mu\nu}v^{\alpha\beta} + a^{\mu\nu}a^{\alpha\beta} \\ &-8i(a^\mu a^\nu v^{\alpha\beta} + a^\mu v^{\nu\alpha}a^\beta + v^{\mu\nu}a^\alpha a^\beta) \\ &-32a^\mu a^\nu a^\alpha a^\beta\Big)\Big), \end{aligned} \tag{6a.1.19}$$

with

$$\begin{aligned} v^{\mu\nu} &= \partial^\mu v^\nu - \partial^\nu v^\mu + i\Big[v^\mu,\, v^\nu\Big] + i\Big[a^\mu,\, a^\nu\Big], \\ a^{\mu\nu} &= \partial^\mu a^\nu - \partial^\nu\, a^\mu + i\Big[v^\mu,\, a^\nu\Big] + i\Big[a^\mu,\, v^\nu\Big]. \end{aligned} \tag{6a.1.20}$$

This is Bardeen's form for the non-Abelian anomaly. We will give a more elementary derivation of parts of this, using Feynman graphs, later in this chapter, and will show you how to derive the rest. There are other forms that are not what we want because they do not satisfy ordinary $SU(3)$ gauge invariance, but they can be obtained from (6a.1.19) with the addition of polynomials in v^μ and a^μ to the action. These correspond to different schemes for renormalizing the underlying theory. However, there is no way of removing the anomaly entirely by such manipulations.

Now we can integrate the anomaly to obtain W. Define

$$\begin{aligned} &U(s) = e^{2is\underset{\sim}{\Pi}/f}, \quad \xi(s) = e^{is\underset{\sim}{\Pi}/f}, \\ &l^\mu(s) = \xi^\dagger(1-s)l^\mu\xi(1-s) - i\xi^\dagger(1-s)\partial^\mu\xi(1-s), \\ &r^\mu(s) = \xi(1-s)r^\mu\xi^\dagger(1-s) - i\xi(1-s)\partial^\mu\xi^\dagger(1-s), \\ &v^\mu(s) = \Big(l^\mu(s) + r^\mu(s)\Big)/2, \\ &a^\mu(s) = \Big(l^\mu(s) - r^\mu(s)\Big)/2. \end{aligned} \tag{6a.1.21}$$

The point of (6a.1.21) is that the different values of s are related by chiral gauge transformations, which depend on the Goldstone boson field $\underset{\sim}{\Pi}$. In particular

$$\begin{aligned} &\frac{d}{ds}W\Big(U(s),\, v^\mu(s),\, a^\mu(s)\Big)\, ds \\ &= \delta_{ds\underset{\sim}{\Pi}/f}W = ds\; F\Big(\underset{\sim}{\Pi}/f,\, v^\mu(s),\, a^\mu(s)\Big). \end{aligned} \tag{6a.1.22}$$

This we can integrate to obtain

$$\begin{aligned} W\big(U,\, v^\mu,\, a^\mu\big) &= W\big(U(1),\, v^\mu(1),\, a^\mu(1)\big) \\ &= \int_0^1 F\big(\underset{\sim}{\Pi}/f,\, v^\mu(s),\, a^\mu(s)\big)\, ds - W\big(U(0),\, v^\mu(0), a^\mu(0)\big). \end{aligned} \tag{6a.1.23}$$

The second term on the right-hand side is actually an $SU(3)_L \times SU(3)_R$ invariant. To see this, note that

$$\begin{aligned} U(0) &= 1 \\ l^\mu(0) &= \xi^\dagger l^\mu \xi - i\xi^\dagger \partial^\mu \xi \\ r^\mu(0) &= \xi r^\mu \xi^\dagger - i\xi \partial^\mu \xi^\dagger, \end{aligned} \tag{6a.1.24}$$

where

$$\xi^2 = U. \tag{6a.1.25}$$

A local $SU(3)_L \times SU(3)_R$ transformation on U defines a local $SU(3)$ transformation u on the ξ's as follows:

$$\begin{aligned} U &\to LUR^\dagger \\ \xi &\to L\xi u^\dagger = u\xi R^\dagger\,. \end{aligned} \tag{6a.1.26}$$

This defines the u matrix as a local $SU(3)$ transformation that depends on L, R, and ξ. From (6a.1.24)–(6a.1.26) it follows that

$$\begin{aligned} l^\mu(0) &\to ul^\mu(0)u^\dagger \quad -u\partial^\mu u^\dagger \\ r^\mu(0) &\to ur^\mu(0)u^\dagger \quad -iu\partial^\mu u^\dagger. \end{aligned} \tag{6a.1.27}$$

But (6a.1.27) has the form of an ordinary $SU(3)$ gauge transformation, and $W\big((1,\, l^\mu(0),\, v^\mu(0)\big)$ is invariant under ordinary $SU(3)$ gauge transformations. Thus, it can be absorbed into the ordinary terms in the action that are invariant under $SU(3)_L \times SU(3)_R$. Hence, we can write the extra term that incorporates the effect of the anomaly as

$$W\big(U,\, v^\mu,\, a^\mu\big) = \int_0^1 F\big(\underset{\sim}{\Pi},\, v^\mu(s),\, a^\mu(s)\big)\, ds\,. \tag{6a.1.28}$$

Now that we have solved the problem (at least formally) for a general $SU(3)_L \times SU(3)_R$ gauge theory, we can specialize to the case in which we are actually interested by setting

$$v^\mu = eQA^\mu, \quad a^\mu = 0 \tag{6a.1.29}$$

in (6a.1.21) and (6a.1.28) to obtain the effect of the electromagnetic anomaly.

Specific interaction terms (such as (6a.1.5) can be obtained by expanding $\xi(s)$ in a power series in $\underset{\sim}{\Pi}$ before doing the s integration.

That was a lot of work, but the result is rather remarkable. Not only does it answer the question that we posed originally, but several others. It gives a term of the form(6a.1.5) (see Problem 5–6), but also a term which produces a $\gamma \to 3\pi$ transition. Even when the gauge field is turned off completely, an effect of the anomaly remains like the Cheshire cat's smile. The leading operator that does not involve the gauge field is

$$\frac{2}{5\pi^2 f^5}\epsilon_{\mu\nu\alpha\beta}\text{tr}\left(\underset{\sim}{\Pi}\partial^\mu\underset{\sim}{\Pi}\partial^\nu\underset{\sim}{\Pi}\partial^\alpha\underset{\sim}{\Pi}\partial^\beta\underset{\sim}{\Pi}\right). \tag{6a.1.30}$$

This term is interesting because it is the leading term in the effective Lagrangian that requires the Goldstone bosons to be pseudoscalars (note that it is absent in $SU(3)_L \times SU(3)_R$ because of G parity). No such term with only four derivatives can be built simply as an $SU(3)_L \times SU(3)_R$ invariant function of the U's.

There is a subtlety hidden in the Wess-Zumino analysis, Our expression for the extra term seems to depend on our ability to define the Goldstone boson field $\underset{\sim}{\Pi}$. But in general, the U field does not uniquely determine the $\underset{\sim}{\Pi}$ field. In perturbation theory, this doesn't bother us. We always assume that the $\underset{\sim}{\Pi}$ field is "small". In fact, E. Witten (*Nuclear Physics* **B223**, 422–432, 1983) has shown that the analysis makes sense even for large fields, provided the coefficient of the anomaly (and therefore of the extra term in the action) is properly quantized. His analysis clarifies the significance of the Wess-Zumino term.

6a.2 The Steinberger Calculation

Jack Steinberger is an outstanding experimental physicist. He did one important piece of **theoretical** physics. He considered a field theory in which pions are coupled to massive fermions (proton and neutron at the time, although they could just as well have been quarks) with the interaction term

$$-\,ig\overline{\psi}\,\gamma_5\,T_a\pi_a\,\psi\,. \tag{6a.2.1}$$

Steinberger calculated the contribution of a fermion loop to the decay $\pi^0 \to \gamma\gamma$ — from the diagram

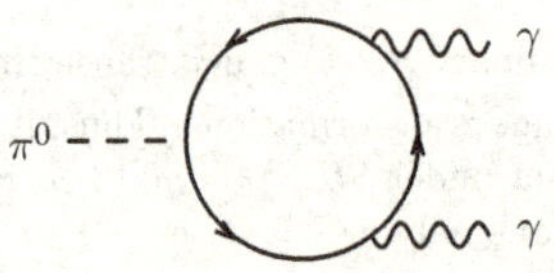

Figure 6a-1:

He found a result that (for the case of three colors of quarks) reproduces the result of the Wess-Zumino calculation, (6a.1.28) (see also problem 6a-1). This has caused some confusion. If the $\pi^0 \to \gamma\gamma$ decay can be produced either by the anomaly or by a quark loop, what happens in the chiral quark model? And for that matter, what about baryon loops? In particular, what does this do to the discussion at the beginning of chapter 6 of how the baryons transform.

Thus we must answer the following question. When should we include the Wess-Zumino term? The way to think about it is to note that the term always appears when the quarks are integrated out of the theory to produce a low-energy theory of the Goldstone boson fields alone. What happens when the quarks are left in the theory as in the chiral quark model depends on how the quarks transform under the chiral symmetry. If the quarks transform nonlinearly as discussed in above, so that all their interactions **explicitly** involve derivatives, then the Wess-Zumino term must be there, just as in the case when the quarks are integrated out. The reason is simply that in this case, the quark interactions go away at low momentum, so quark loops are harmless at low momenta — they do not produce any new interactions. We can figure out what happens for other transformation laws by actually building the fields with the desired transformation laws using the Goldstone boson fields.

For example, suppose we want to from quark fields, q, transforming nonlinearly under $SU(3) \times SU(3)$, to quark fields, q', transforming linearly as

$$q'_L \to L\, q'_L\,, \qquad q'_R \to R\, q'_R\,. \tag{6a.2.2}$$

The q' quarks are related to the q quarks by a chiral transformation

$$q'_L = \xi\, q_L\,, \qquad q'_R = \xi^\dagger\,. \tag{6a.2.3}$$

In terms of the q' quarks, the mass term becomes

$$-m\bar{q}q \to -m\overline{q'_R}U^\dagger q'_L - m\overline{q_L}U q_R \tag{6a.2.4}$$

which contains nonderivative couplings of the Goldstone bosons to the quarks. You can check that the couplings are determined by the Goldberger-Trieman relation, in its **naive** form, (2.6.25) (with no factor of g_A).

But we saw in the previous section that the Wess-Zumino term is precisely the response of the Lagrangian to such a chiral transformation. The transformation (6a.2.3) precisely cancels the Wess-Zumino term. Thus we expect the Steinberger calculation to reproduce the effects of the Wess-Zumino term when the nonderivative coupling is given by (6a.2.4).

6a.3 Spectators, gauge invariance and the anomaly

Here is a diagramatic way to determine the form of the anomaly. It is really just another way of seeing that the Steinberger calculation must give the right answer, but is perhaps a little cleaner and more convincing than the argument we gave in the previous section. Suppose that we add to the QCD theory of three massless flavors, three sets of three spectator fermions, to cancel the chiral flavor anomalies. Specifically, add to the theory of colored quarks,

$$q = \begin{pmatrix} u \\ d \\ s \end{pmatrix} \tag{6a.3.1}$$

a set of three massless spectator fermions

$$Q^j = \begin{pmatrix} U^j \\ U^j \\ U^j \end{pmatrix} \quad \text{for } j = 1 \text{ to } 3. \tag{6a.3.2}$$

These massless fermions have their own chiral symmetries, but we can break the symmetry down to $SU(3) \times SU(3)$ with a four-fermion interaction of the following form:

$$\frac{1}{\Lambda^2}\left(\overline{q_R}\, Q^j_L\right)\left(\overline{Q^j_R}\, q_L\right) + \text{h.c.} \tag{6a.3.3}$$

This term ensures that q_L and Q^j_R rotate under the same $SU(3)_L$, and that q_R and Q^j_L rotate under the same $SU(3)_R$. It is not renormalizable, but that doesn't matter to our argument. We could generate such a term by exchange of a scalar field, but since we will only use it a low energy scales, it doesn't really matter where it comes from. Notice also that this term is constructed to be a color singlet, and also a singlet under a global $SU(3)$ acting on the j index (this global symmetry is unneccesary, but simplifies the tableaux).

Now there are no anomalies to break the chiral flavor symmetries, so if we couple in classical flavor gauge fields, as usual, the theory is guage invariant. Now QCD gets strong and the chiral symmetry is spontaneously broken, and realized nonlinearly.

Now what happens to the term (6a.3.3) in the low energy theory? This is actually easy to see from (5.3.1) and (5.4.3). The point is that because the spectator fermions do not carry the color $SU(3)$, the term

$$\frac{1}{\Lambda^2} Q_L^j \overline{Q_R^j} \tag{6a.3.4}$$

enters into the theory in exactly the same way as $s+ip$ in (5.3.1). Thus in the low energy theory, it also must appear in the same way as $s+ip$. The leading term a low energies that does not involve derivatives of Π is therefore one in which the term (6a.3.4) appears in the same way as $s+ip$ appears in (5.4.3). Thus the term is

$$M \overline{Q_R^j} U Q_L^j + \text{h.c.} \tag{6a.3.5}$$

where $M = v^3/\Lambda^2$. This is a spectator mass term in the low energy theory, along with a specific set of Goldstone bosons couplings consistent with the Goldberger-Treiman relation.

Now as we go down further down in energy scale, below the spectator mass, we must get a gauge invariant theory involving only Goldstone bosons. Spectator loops must cancel any gauge variant effects in the low energy chiral theory. Thus the spectator loops must give minus the WZ term. This has the virtue that we can regularize it to preserve the vector symmetry, and that should give the conventional WZ term. We can now take the spectator mass, M, to infinity, because the theory has to make sense at arbitrarily long distances, so the only things that can matter are the terms that are independent of the mass.

This analysis shows that we can get any coupling in the Wess-Zumino term from a Steinberger-like loop calculation. But we can also just calculate the coupling of a single Goldstone boson to the external gauge fields. This is actually equivalent to finding the result of an infinitesimal chiral transformation, because that is what the Goldstone boson field represents. Thus this should reproduce the Wess-Zumino analysis that leads to Bardeen form for the anomaly. The we can get the rest of the Wess-Zumino term by integrating the anomaly, as we did in the first section. What graphs can contribute? The Goldstone boson coupling is proportional to $1/f$. Thus to get a contribution that is independent of M, we need a dimension five operators. Thus only the three-, four- and five-point functions can contribute on dimensional grounds (any operator with dimension higher than five will be suppressed by powers M). There are only a few possibilities that are proportional to $\epsilon^{\mu\nu\alpha\beta}$ and consistent with the vector $SU(3)$, parity and Lorentz invariance. They are

$$\begin{gathered}\epsilon^{\mu\nu\alpha\beta}\operatorname{tr}\left(\Pi V_{\mu\nu} V_{\alpha\beta}\right), \quad \epsilon^{\mu\nu\alpha\beta}\operatorname{tr}\left(\Pi A_{\mu\nu} A_{\alpha\beta}\right), \quad \epsilon^{\mu\nu\alpha\beta}\operatorname{tr}\left(\Pi V_{\mu\nu} a_\alpha a_\beta\right), \\ \epsilon^{\mu\nu\alpha\beta}\operatorname{tr}\left(\Pi a_\mu V_{\nu\alpha} a_\beta\right), \quad \epsilon^{\mu\nu\alpha\beta}\operatorname{tr}\left(\Pi a_\mu a_\nu V_{\alpha\beta}\right), \quad \epsilon^{\mu\nu\alpha\beta}\operatorname{tr}\left(\Pi a_\mu a_\nu a_\alpha a_\beta\right),\end{gathered} \tag{6a.3.6}$$

where

$$V_{\mu\nu} = \partial_\mu v_\nu - \partial_\nu v_\mu + i\left[v_\mu, v_\nu\right], \qquad A_{\mu\nu} = \partial_\mu a_\nu - \partial_\nu a_\mu + i\left[v_\mu, a_\nu\right] + i\left[a_\mu, v_\nu\right]. \tag{6a.3.7}$$

In terms of the these objects, the Bardeen anomaly, (6a.1.19), with $\underset{\sim}{c} \to \Pi$ becomes

$$
\begin{aligned}
f = -\frac{1}{16\pi^2}\epsilon_{\mu\nu\alpha\beta}\,\mathrm{tr}\Big(\Pi\Big(&3V^{\mu\nu}V^{\alpha\beta} + A^{\mu\nu}A^{\alpha\beta} \\
&-2i(a^\mu a^\nu V^{\alpha\beta} + V^{\mu\nu}a^\alpha a^\beta) - 8ia^\mu V^{\nu\alpha}a^\beta + \\
&+4a^\mu a^\nu a^\alpha a^\beta\Big)\Big),
\end{aligned}
\tag{6a.3.8}
$$

Below, we will verify each of these terms.

Let us start by calculating the terms proportional to two field-strengths, the first two terms in (6a.3.6). To do that we consider the three point function for $SU(3)\times SU(3)$ vector and axial currents and a Goldstone field.

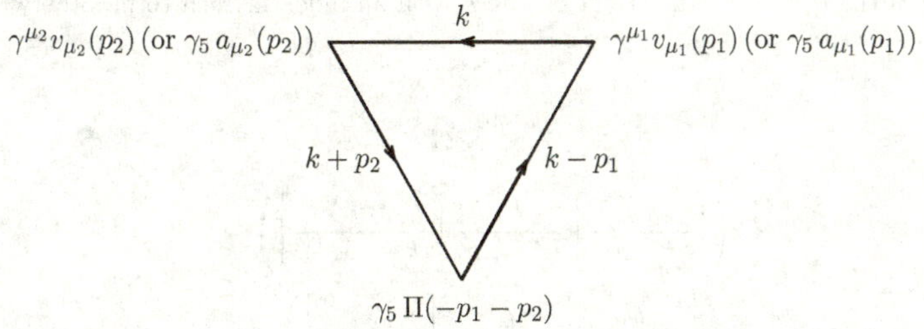

(6a.3.9)

The ΠVV term is

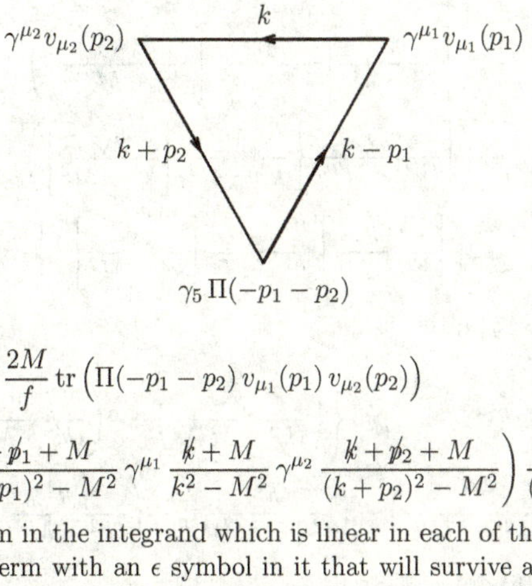

(6a.3.10)

$$
\begin{gathered}
\frac{2M}{f}\,\mathrm{tr}\Big(\Pi(-p_1-p_2)\,v_{\mu_1}(p_1)\,v_{\mu_2}(p_2)\Big) \\
\int \mathrm{tr}\left(\gamma_5\,\frac{\not k - \not p_1 + M}{(k-p_1)^2-M^2}\,\gamma^{\mu_1}\,\frac{\not k + M}{k^2-M^2}\,\gamma^{\mu_2}\,\frac{\not k + \not p_2 + M}{(k+p_2)^2-M^2}\right)\frac{d^4k}{(2\pi)^4}
\end{gathered}
\tag{6a.3.11}
$$

We are interested in a term in the integrand which is linear in each of the momenta, because there is no other way to get a term with an ϵ symbol in it that will survive as $M\to\infty$. Expanding in powers of momentum and keeping only the relevant term, we can write the integrand as

$$
-\,\mathrm{tr}\left(\gamma_5\,p_1^\nu\left[\frac{\partial}{\partial k^\nu}\frac{\not k + M}{k^2-M^2}\right]\gamma^{\mu_1}\,\frac{\not k + M}{k^2-M^2}\,\gamma^{\mu_2}\,p_2^\lambda\left[\frac{\partial}{\partial k^\lambda}\frac{\not k + M}{k^2-M^2}\right]\right)
\tag{6a.3.12}
$$

We can do the trace before performing the differentiations. One of the propagators must contribute an M and the two others must contribute γs. Using

$$
\mathrm{tr}\left(\gamma_5\gamma^\mu\gamma^\nu\gamma^\alpha\gamma^\beta\right) = 4i\,\epsilon^{\mu\nu\alpha\beta}
\tag{6a.3.13}
$$

we have

$$-4i\,\epsilon^{\mu_1\mu_2\alpha\beta}\,p_1^\nu\,p_2^\lambda$$

$$\begin{aligned}&\left\{-\left(\left[\frac{\partial}{\partial k^\nu}\frac{k_\alpha}{k^2-M^2}\right]\frac{k_\beta}{k^2-M^2}\left[\frac{\partial}{\partial k^\lambda}\frac{M}{k^2-M^2}\right]\right)\right.\\&+\left(\left[\frac{\partial}{\partial k^\nu}\frac{k_\alpha}{k^2-M^2}\right]\frac{M}{k^2-M^2}\left[\frac{\partial}{\partial k^\lambda}\frac{k_\beta}{k^2-M^2}\right]\right)\\&\left.-\left(\left[\frac{\partial}{\partial k^\nu}\frac{M}{k^2-M^2}\right]\frac{k_\alpha}{k^2-M^2}\left[\frac{\partial}{\partial k^\lambda}\frac{k_\beta}{k^2-M^2}\right]\right)\right\}\end{aligned} \tag{6a.3.14}$$

At least one of the derivatives must act on the k with an index in each term, otherwise the ϵ vanishes, so we can write this as

$$-4i\,\epsilon^{\mu_1\mu_2\alpha\beta}\,p_1^\nu\,p_2^\lambda$$

$$\begin{aligned}&\left\{2\left(\left[\frac{g_{\nu\alpha}}{k^2-M^2}\right]\frac{k_\beta}{k^2-M^2}\left[\frac{M\,k^\lambda}{(k^2-M^2)^2}\right]\right)\right.\\&+\left(\left[\frac{g_{\nu\alpha}}{k^2-M^2}\right]\frac{M}{k^2-M^2}\left[\frac{g_{\lambda\beta}}{k^2-M^2}\right]\right)\\&-2\left(\left[\frac{g_{\nu\alpha}}{k^2-M^2}\right]\frac{M}{k^2-M^2}\left[\frac{k_\lambda\,k_\beta}{(k^2-M^2)^2}\right]\right)\\&-2\left(\left[\frac{k_\nu\,k_\alpha}{(k^2-M^2)^2}\right]\frac{M}{k^2-M^2}\left[\frac{g_{\lambda\beta}}{k^2-M^2}\right]\right)\\&\left.+2\left(\left[\frac{M\,k_\nu}{k^2-M^2}\right]\frac{k_\alpha}{k^2-M^2}\left[\frac{g_{\lambda\beta}}{k^2-M^2}\right]\right)\right\}\end{aligned} \tag{6a.3.15}$$

Now since $2\,k_\alpha\,k_\beta \to \frac{1}{2}k^2\,g_{\alpha\beta}$, this is

$$\begin{aligned}&-2i\,M\,\epsilon^{\mu_1\mu_2\alpha\beta}\,p_{1\alpha}\,p_{2\beta}\left\{\left(\frac{k^2}{(k^2-M^2)^4}\right)+2\left(\frac{1}{(k^2-M^2)^3}\right)\right.\\&\left.-\left(\frac{k^2}{(k^2-M^2)^4}\right)-\left(\frac{k^2}{(k^2-M^2)^4}\right)+\left(\frac{k^2}{(k^2-M^2)^4}\right)\right\}\end{aligned} \tag{6a.3.16}$$

The integrals in (6a.3.16) are

$$\int\frac{1}{(k^2-M^2)^3}\frac{d^4k}{(2\pi)^4}=-i\frac{1}{32\pi^2\,M^2}\qquad\int\frac{k^2}{(k^2-M^2)^4}\frac{d^4k}{(2\pi)^4}=-i\frac{1}{48\pi^2\,M^2} \tag{6a.3.17}$$

Using this and putting in the factors from (6a.3.11), and multiplying by three, because we need three different kinds of spectators because there are three colors of quarks, we get, in momentum space,

$$-\frac{3}{4\pi^2 f}\,\mathrm{tr}\left(\Pi(-p_1-p_2)\,v_{\mu_1}(p_1)\,v_{\mu_2}(p_2)\right)\epsilon^{\mu_1\mu_2\alpha\beta}\,p_{1\alpha}\,p_{2\beta} \tag{6a.3.18}$$

which corresponds to

$$-\frac{3}{16\pi^2 f}\,\epsilon^{\mu\nu\alpha\beta}\,\mathrm{tr}\left(\Pi\, V_{\mu\nu}\, V_{\alpha\beta}\right). \tag{6a.3.19}$$

The calculation of the term with two axial vector currents proceeds similarly. The relevant integral is

$$\begin{gathered}\frac{2M}{f}\,\mathrm{tr}\left(\Pi(-p_1-p_2)\,a_{\mu_1}(p_1)\,a_{\mu_2}(p_2)\right)\\ \int \mathrm{tr}\left(\gamma_5\,\frac{\not k-\not p_1+M}{(k-p_1)^2-M^2}\,\gamma^{\mu_1}\gamma_5\,\frac{\not k+M}{k^2-M^2}\,\gamma^{\mu_2}\gamma_5\,\frac{\not k+\not p_2+M}{(k+p_2)^2-M^2}\right)\frac{d^4k}{(2\pi)^4}\end{gathered} \tag{6a.3.20}$$

We can proceed as before. In fact, we can immediately turn (6a.3.20) into something very like (6a.3.12) by moving the γ_5s around —

$$-\,\mathrm{tr}\left(\gamma_5\,p_1^\nu\left[\frac{\partial}{\partial k^\nu}\frac{\not k+M}{k^2-M^2}\right]\gamma^{\mu_1}\,\frac{\not k-M}{k^2-M^2}\,\gamma^{\mu_2}\,p_2^\lambda\left[\frac{\partial}{\partial k^\lambda}\frac{\not k+M}{k^2-M^2}\right]\right) \tag{6a.3.21}$$

The only difference between (6a.3.21) and (6a.3.12) is the sign of the M in the middle propagator. Thus the terms proportional to this factor of M just change sign:

$$\begin{gathered}-4i\,\epsilon^{\mu_1\mu_2\alpha\beta}\,p_1^\nu\,p_2^\lambda\\ \left\{-\left(\left[\frac{\partial}{\partial k^\nu}\frac{k_\alpha}{k^2-M^2}\right]\frac{k_\beta}{k^2-M^2}\left[\frac{\partial}{\partial k^\lambda}\frac{M}{k^2-M^2}\right]\right)\right.\\ -\left(\left[\frac{\partial}{\partial k^\nu}\frac{k_\alpha}{k^2-M^2}\right]\frac{M}{k^2-M^2}\left[\frac{\partial}{\partial k^\lambda}\frac{k_\beta}{k^2-M^2}\right]\right)\\ \left.-\left(\left[\frac{\partial}{\partial k^\nu}\frac{M}{k^2-M^2}\right]\frac{k_\alpha}{k^2-M^2}\left[\frac{\partial}{\partial k^\lambda}\frac{k_\beta}{k^2-M^2}\right]\right)\right\}\end{gathered} \tag{6a.3.22}$$

which we can write this as

$$\begin{gathered}-4i\,\epsilon^{\mu_1\mu_2\alpha\beta}\,p_1^\nu\,p_2^\lambda\\ \left\{2\left(\left[\frac{g_{\nu\alpha}}{k^2-M^2}\right]\frac{k_\beta}{k^2-M^2}\left[\frac{M\,k^\lambda}{(k^2-M^2)^2}\right]\right)\right.\\ -\left(\left[\frac{g_{\nu\alpha}}{k^2-M^2}\right]\frac{M}{k^2-M^2}\left[\frac{g_{\lambda\beta}}{k^2-M^2}\right]\right)\\ +2\left(\left[\frac{g_{\nu\alpha}}{k^2-M^2}\right]\frac{M}{k^2-M^2}\left[\frac{k_\lambda\,k_\beta}{(k^2-M^2)^2}\right]\right)\\ +2\left(\left[\frac{k_\nu\,k_\alpha}{(k^2-M^2)^2}\right]\frac{M}{k^2-M^2}\left[\frac{g_{\lambda\beta}}{k^2-M^2}\right]\right)\\ \left.+2\left(\left[\frac{M\,k_\nu}{k^2-M^2}\right]\frac{k_\alpha}{k^2-M^2}\left[\frac{g_{\lambda\beta}}{k^2-M^2}\right]\right)\right\}\end{gathered} \tag{6a.3.23}$$

Now since $2\,k_\alpha\,k_\beta \to \frac{1}{2}k^2\,g_{\alpha\beta}$, this is

$$\begin{aligned}&-2i\,M\,\epsilon^{\mu_1\mu_2\alpha\beta}\,p_{1\alpha}\,p_{2\beta}\left\{\left(\frac{k^2}{(k^2-M^2)^4}\right)-2\left(\frac{1}{(k^2-M^2)^3}\right)\right.\\&\left.+\left(\frac{k^2}{(k^2-M^2)^4}\right)+\left(\frac{k^2}{(k^2-M^2)^4}\right)+\left(\frac{k^2}{(k^2-M^2)^4}\right)\right\}\end{aligned} \tag{6a.3.24}$$

Thus the result is

$$-\frac{1}{4\pi^2 f}\,\mathrm{tr}\left(\Pi(-p_1-p_2)\,a_{\mu_1}(p_1)\,a_{\mu_2}(p_2)\right)\epsilon^{\mu_1\mu_2\alpha\beta}\,p_{1\alpha}\,p_{2\beta} \tag{6a.3.25}$$

corresponding to

$$-\frac{1}{16\pi^2 f}\,\epsilon^{\mu\nu\alpha\beta}\,\mathrm{tr}\left(\Pi\,A_{\mu\nu}\,A_{\alpha\beta}\right). \tag{6a.3.26}$$

Now let us look for the terms proportional to one field-strength. For that we need to look at four-point functions. We could do all of these at once by putting in a general momentum dependence, but it is a little simpler to keep track of what is going on if we restrict ourselves to external momentum dependence that picks out a single term in (6a.3.6). We can do this by having the external momentum carried only by the Π and v fields, because we know from (6a.3.6) that if picked out terms proportional to the momenta of the a fields, the terms would either vanish or reproduce what we already calculated in (6a.3.26). So for example, we look at the following graph:

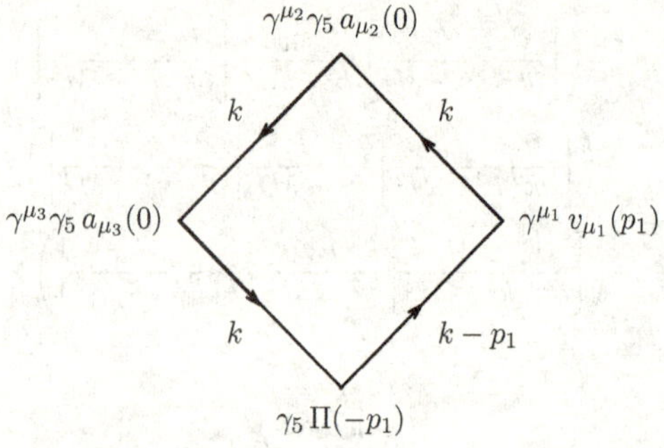

(6a.3.27)

The Feynman integral is

$$\begin{aligned}&\frac{2M}{f}\,\mathrm{tr}\left(\Pi(-p_1)\,v_{\mu_1}(p_1)\,a_{\mu_2}(0)\,a_{\mu_3}(0)\right)\\&\int \mathrm{tr}\left(\gamma^{\mu_3}\gamma_5\,\frac{\not k+M}{k^2-M^2}\,\gamma_5\,\frac{\not k-\not p_1+M}{(k-p_1)^2-M^2}\,\gamma^{\mu_1}\,\frac{\not k+M}{k^2-M^2}\,\gamma^{\mu_2}\gamma_5\,\frac{\not k+M}{k^2-M^2}\right)\frac{d^4k}{(2\pi)^4}\end{aligned} \tag{6a.3.28}$$

Begin by writing the integrand as follows:

$$\begin{aligned}&-\mathrm{tr}\left(\gamma^{\mu_3}\,\frac{1}{(k-p_1)^2-M^2}\,\gamma^{\mu_1}\,\frac{\not k+M}{k^2-M^2}\,\gamma^{\mu_2}\gamma_5\,\frac{\not k+M}{k^2-M^2}\right)\\&+\mathrm{tr}\left(\gamma^{\mu_3}\,\frac{\not k-M}{k^2-M^2}\,\frac{\not p_1}{(k-p_1)^2-M^2}\,\gamma^{\mu_1}\,\frac{\not k+M}{k^2-M^2}\,\gamma^{\mu_2}\gamma_5\,\frac{\not k+M}{k^2-M^2}\right)\end{aligned} \tag{6a.3.29}$$

In the second term, because the numerator is explicitly proportional to p_1 and since we are looking for a term involving only one power of p_1 we can drop the p_1 dependence in denominator. In the first term, we must expand in powers of p_1 and pick out the first term:

$$
\begin{aligned}
&-\mathrm{tr}\left(\gamma^{\mu_3}\frac{2k^\mu p_{1\mu}}{(k^2-M^2)^2}\gamma^{\mu_1}\frac{\not{k}+M}{k^2-M^2}\gamma^{\mu_2}\gamma_5\frac{\not{k}+M}{k^2-M^2}\right)\\
&+\mathrm{tr}\left(\gamma^{\mu_3}\frac{\not{k}-M}{k^2-M^2}\frac{\not{p}_1}{k^2-M^2}\gamma^{\mu_1}\frac{\not{k}+M}{k^2-M^2}\gamma^{\mu_2}\gamma_5\frac{\not{k}+M}{k^2-M^2}\right)
\end{aligned}
\tag{6a.3.30}
$$

We now average over the k directions, replacing $k^\mu k^\nu \to \frac{1}{4}g^{\mu\nu}$ and dropping terms odd in k to obtain

$$
\begin{aligned}
&-\frac{1}{2}k^2\,\mathrm{tr}\left(\gamma^{\mu_3}\frac{1}{(k^2-M^2)^2}\gamma^{\mu_1}\frac{\not{p}_1}{k^2-M^2}\gamma^{\mu_2}\gamma_5\frac{M}{k^2-M^2}\right)\\
&-\frac{1}{2}k^2\,\mathrm{tr}\left(\gamma^{\mu_3}\frac{1}{(k^2-M^2)^2}\gamma^{\mu_1}\frac{M}{k^2-M^2}\gamma^{\mu_2}\gamma_5\frac{\not{p}_1}{k^2-M^2}\right)\\
&+\frac{1}{4}k^2\,\mathrm{tr}\left(\gamma^{\mu_3}\frac{\gamma_\mu}{k^2-M^2}\frac{\not{p}_1}{k^2-M^2}\gamma^{\mu_1}\frac{\gamma^\mu}{k^2-M^2}\gamma^{\mu_2}\gamma_5\frac{M}{k^2-M^2}\right)\\
&+\frac{1}{4}k^2\,\mathrm{tr}\left(\gamma^{\mu_3}\frac{\gamma_\mu}{k^2-M^2}\frac{\not{p}_1}{k^2-M^2}\gamma^{\mu_1}\frac{M}{k^2-M^2}\gamma^{\mu_2}\gamma_5\frac{\gamma^\mu}{k^2-M^2}\right)\\
&-\frac{1}{4}k^2\,\mathrm{tr}\left(\gamma^{\mu_3}\frac{M}{k^2-M^2}\frac{\not{p}_1}{k^2-M^2}\gamma^{\mu_1}\frac{\gamma_\mu}{k^2-M^2}\gamma^{\mu_2}\gamma_5\frac{\gamma^\mu}{k^2-M^2}\right)\\
&-\mathrm{tr}\left(\gamma^{\mu_3}\frac{M}{k^2-M^2}\frac{\not{p}_1}{k^2-M^2}\gamma^{\mu_1}\frac{M}{k^2-M^2}\gamma^{\mu_2}\gamma_5\frac{M}{k^2-M^2}\right)
\end{aligned}
\tag{6a.3.31}
$$

In the third through fifth terms in (6a.3.31), we can use the standard identities,

$$
\gamma^\mu\gamma^\alpha\gamma^\beta\gamma_\mu = 4g^{\alpha\beta}\,,\qquad \gamma^\mu\gamma^\alpha\gamma_\mu = -2\gamma^\alpha\,,\qquad \gamma^\mu\gamma^\alpha\gamma_5\gamma_\mu = 2\gamma^\alpha\gamma_5\,,
\tag{6a.3.32}
$$

to get rid of pairs of γs, to obtain

$$
\begin{aligned}
&-\frac{1}{2}k^2\,\mathrm{tr}\left(\gamma^{\mu_3}\,\frac{1}{(k^2-M^2)^2}\,\gamma^{\mu_1}\,\frac{\not{p}_1}{k^2-M^2}\,\gamma^{\mu_2}\gamma_5\,\frac{M}{k^2-M^2}\right)\\
&-\frac{1}{2}k^2\,\mathrm{tr}\left(\gamma^{\mu_3}\,\frac{1}{(k^2-M^2)^2}\,\gamma^{\mu_1}\,\frac{M}{k^2-M^2}\,\gamma^{\mu_2}\gamma_5\,\frac{\not{p}_1}{k^2-M^2}\right)\\
&+k^2\,\mathrm{tr}\left(\gamma^{\mu_3}\,\frac{p_1{}^{\mu_1}}{(k^2-M^2)^2}\,\gamma^{\mu_2}\gamma_5\,\frac{M}{k^2-M^2}\right)\\
&-\frac{1}{2}k^2\,\mathrm{tr}\left(\gamma^{\mu_3}\,\frac{1}{k^2-M^2}\,\frac{\not{p}_1}{k^2-M^2}\,\gamma^{\mu_1}\,\frac{M}{k^2-M^2}\,\gamma^{\mu_2}\gamma_5\,\frac{1}{k^2-M^2}\right)\\
&-\frac{1}{2}k^2\,\mathrm{tr}\left(\gamma^{\mu_3}\,\frac{M}{k^2-M^2}\,\frac{\not{p}_1}{k^2-M^2}\,\gamma^{\mu_1}\,\frac{1}{k^2-M^2}\,\gamma^{\mu_2}\gamma_5\,\frac{1}{k^2-M^2}\right)\\
&-\mathrm{tr}\left(\gamma^{\mu_3}\,\frac{M}{k^2-M^2}\,\frac{\not{p}_1}{k^2-M^2}\,\gamma^{\mu_1}\,\frac{M}{k^2-M^2}\,\gamma^{\mu_2}\gamma_5\,\frac{M}{k^2-M^2}\right)
\end{aligned}
\tag{6a.3.33}
$$

The third term vanishes. In the rest, we can use (6a.3.13) to obtain

$$
\begin{aligned}
&4i\,\epsilon^{\mu\mu_1\mu_2\mu_3}\,p_{1\mu}\left[\frac{M\,k^2}{(k^2-M^2)^4}\left\{-\frac{1}{2}-\frac{1}{2}+\frac{1}{2}+\frac{1}{2}\right\}+\frac{M^3}{(k^2-M^2)^4}\right]\\
&\qquad = 4i\,\epsilon^{\mu\mu_1\mu_2\mu_3}\,p_{1\mu}\,\frac{M^3}{(k^2-M^2)^4}\,.
\end{aligned}
\tag{6a.3.34}
$$

Now using

$$
\int \frac{1}{(k^2-M^2)^4}\,\frac{d^4k}{(2\pi)^4} = i\,\frac{1}{96\pi^2\,M^4}
\tag{6a.3.35}
$$

we get (putting in the color factor)

$$
-\frac{1}{4\pi^2 f}\,\mathrm{tr}\left(\Pi(-p_1)\,p_{1\mu}v_{\mu_1}(p_1)\,a_{\mu_2}(0)\,a_{\mu_3}(0)\right)\epsilon^{\mu\mu_1\mu_2\mu_3}
\tag{6a.3.36}
$$

which corresponds to

$$
\frac{i}{8\pi^2 f}\,\epsilon^{\mu\mu_1\mu_2\mu_3}\,\mathrm{tr}\left(\Pi\,V_{\mu\mu_1}\,a_{\mu_2}\;a_{\mu_3}\,\cdot\right)
\tag{6a.3.37}
$$

Now look at the following graph:

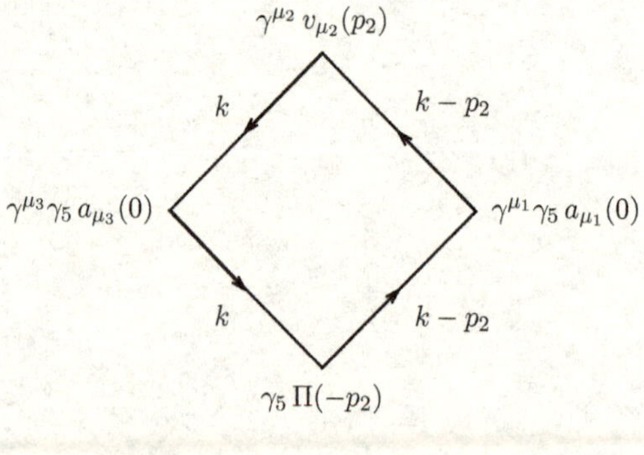

(6a.3.38)

The Feynman integral is

$$\frac{2M}{f}\,\mathrm{tr}\left(\Pi(-p_2)\,a_{\mu_1}(0)\,v_{\mu_2}(p_2)\,a_{\mu_3}(0)\right)$$
$$\int \mathrm{tr}\left(\gamma^{\mu_3}\gamma_5\,\frac{\not k+M}{k^2-M^2}\,\gamma_5\,\frac{\not k-\not p_2+M}{(k-p_2)^2-M^2}\,\gamma^{\mu_1}\gamma_5\,\frac{\not k-\not p_2+M}{(k-p_2)^2-M^2}\,\gamma^{\mu_2}\,\frac{\not k+M}{k^2-M^2}\right)\frac{d^4k}{(2\pi)^4} \tag{6a.3.39}$$

Again, we start by writing the integrand as follows:

$$\begin{aligned}&-\mathrm{tr}\left(\gamma^{\mu_3}\,\frac{1}{(k-p_2)^2-M^2}\,\gamma^{\mu_1}\gamma_5\,\frac{\not k-\not p_2+M}{(k-p_2)^2-M^2}\,\gamma^{\mu_2}\,\frac{\not k+M}{k^2-M^2}\right)\\ &+\mathrm{tr}\left(\gamma^{\mu_3}\,\frac{\not k-M}{k^2-M^2}\,\frac{\not p_2}{(k-p_2)^2-M^2}\,\gamma^{\mu_1}\gamma_5\,\frac{\not k-\not p_2+M}{(k-p_2)^2-M^2}\,\gamma^{\mu_2}\,\frac{\not k+M}{k^2-M^2}\right)\end{aligned} \tag{6a.3.40}$$

Now we expand in powers of p_2 and pick out the first term:

$$\begin{aligned}&-\mathrm{tr}\left(\gamma^{\mu_3}\,\frac{2k^\mu p_{2\mu}}{(k^2-M^2)^2}\,\gamma^{\mu_1}\gamma_5\,\frac{\not k+M}{k^2-M^2}\,\gamma^{\mu_2}\,\frac{\not k+M}{k^2-M^2}\right)\\ &+\mathrm{tr}\left(\gamma^{\mu_3}\,\frac{1}{k^2-M^2}\,\gamma^{\mu_1}\gamma_5\,\frac{\not p_2}{k^2-M^2}\,\gamma^{\mu_2}\,\frac{\not k+M}{k^2-M^2}\right)\\ &-\mathrm{tr}\left(\gamma^{\mu_3}\,\frac{1}{k^2-M^2}\,\gamma^{\mu_1}\gamma_5\,\frac{(\not k+M)\,2k^\mu p_{2\mu}}{(k^2-M^2)^2}\,\gamma^{\mu_2}\,\frac{\not k+M}{k^2-M^2}\right)\\ &+\mathrm{tr}\left(\gamma^{\mu_3}\,\frac{\not k-M}{k^2-M^2}\,\frac{\not p_2}{k^2-M^2}\,\gamma^{\mu_1}\gamma_5\,\frac{\not k+M}{k^2-M^2}\,\gamma^{\mu_2}\,\frac{\not k+M}{k^2-M^2}\right)\end{aligned} \tag{6a.3.41}$$

We again average over the k directions, replacing $k^\mu k^\nu \to \frac{1}{4}g^{\mu\nu}$ and dropping terms odd in k to obtain

$$
\begin{aligned}
&-\frac{1}{2}\,k^2\,\mathrm{tr}\left(\gamma^{\mu_3}\,\frac{1}{(k^2-M^2)^2}\,\gamma^{\mu_1}\gamma_5\,\frac{\not p_2}{k^2-M^2}\,\gamma^{\mu_2}\,\frac{M}{k^2-M^2}\right)\\
&-\frac{1}{2}\,k^2\,\mathrm{tr}\left(\gamma^{\mu_3}\,\frac{1}{(k^2-M^2)^2}\,\gamma^{\mu_1}\gamma_5\,\frac{M}{k^2-M^2}\,\gamma^{\mu_2}\,\frac{\not p_2}{k^2-M^2}\right)\\
&\qquad+\mathrm{tr}\left(\gamma^{\mu_3}\,\frac{1}{k^2-M^2}\,\gamma^{\mu_1}\gamma_5\,\frac{\not p_2}{k^2-M^2}\,\gamma^{\mu_2}\,\frac{M}{k^2-M^2}\right)\\
&-\frac{1}{2}\,k^2\,\mathrm{tr}\left(\gamma^{\mu_3}\,\frac{1}{k^2-M^2}\,\gamma^{\mu_1}\gamma_5\,\frac{\not p_2}{(k^2-M^2)^2}\,\gamma^{\mu_2}\,\frac{M}{k^2-M^2}\right)\\
&-\frac{1}{2}\,k^2\,\mathrm{tr}\left(\gamma^{\mu_3}\,\frac{1}{k^2-M^2}\,\gamma^{\mu_1}\gamma_5\,\frac{M}{(k^2-M^2)^2}\,\gamma^{\mu_2}\,\frac{\not p_2}{k^2-M^2}\right)\\
&-\mathrm{tr}\left(\gamma^{\mu_3}\,\frac{M}{k^2-M^2}\,\frac{\not p_2}{k^2-M^2}\,\gamma^{\mu_1}\gamma_5\,\frac{M}{k^2-M^2}\,\gamma^{\mu_2}\,\frac{M}{k^2-M^2}\right)\\
&+\frac{1}{4}\,k^2\,\mathrm{tr}\left(\gamma^{\mu_3}\,\frac{\gamma_\mu}{k^2-M^2}\,\frac{\not p_2}{k^2-M^2}\,\gamma^{\mu_1}\gamma_5\,\frac{M}{k^2-M^2}\,\gamma^{\mu_2}\,\frac{\gamma^\mu}{k^2-M^2}\right)\\
&-\frac{1}{4}\,k^2\,\mathrm{tr}\left(\gamma^{\mu_3}\,\frac{M}{k^2-M^2}\,\frac{\not p_2}{k^2-M^2}\,\gamma^{\mu_1}\gamma_5\,\frac{\gamma_\mu}{k^2-M^2}\,\gamma^{\mu_2}\,\frac{\gamma^\mu}{k^2-M^2}\right)\\
&+\frac{1}{4}\,k^2\,\mathrm{tr}\left(\gamma^{\mu_3}\,\frac{\gamma_\mu}{k^2-M^2}\,\frac{\not p_2}{k^2-M^2}\,\gamma^{\mu_1}\gamma_5\,\frac{\gamma^\mu}{k^2-M^2}\,\gamma^{\mu_2}\,\frac{M}{k^2-M^2}\right)
\end{aligned}
\tag{6a.3.42}
$$

In the last three terms in (6a.3.42), we can again use (6a.3.32) to get rid of pairs of γs. Then we can perform all the traces (the last term vanishes)

$$
\begin{aligned}
&4i\,\epsilon^{\mu_1\mu\mu_2\mu_3}\,p_{2\mu}\left[\left(\frac{1}{2}-\frac{1}{2}\right)\frac{M\,k^2}{(k^2-M^2)^4}-\frac{M}{(k^2-M^2)^3}\right.\\
&\left.+\left(\frac{1}{2}-\frac{1}{2}\right)\frac{M\,k^2}{(k^2-M^2)^4}+\frac{M^3}{(k^2-M^2)^4}+\left(\frac{1}{2}-\frac{1}{2}\right)\frac{M\,k^2}{(k^2-M^2)^4}\right]
\end{aligned}
\tag{6a.3.43}
$$

Now using (6a.3.17) and (6a.3.35) and putting in the color factor, we get

$$
-\frac{1}{\pi^2 f}\,\mathrm{tr}\left(\Pi(-p_2)\,a_{\mu_1}(0)\,p_{2\mu}v_{\mu_2}(p_2)\,a_{\mu_3}(0)\right)\epsilon^{\mu_1\mu\mu_2\mu_3}\,p_{2\mu}
\tag{6a.3.44}
$$

which corresponds to

$$
\frac{i}{2\pi^2 f}\,\epsilon^{\mu_1\mu\mu_2\mu_3}\,\mathrm{tr}\left(\Pi\,a_{\mu_1}\,.\,V_{\mu\mu_2}\,a_{\mu_3}\right)
\tag{6a.3.45}
$$

Finally, let us consider the five-point function with four a fields, the last of the possible terms in (6a.3.6). Here, there are no derivatives, so we can take all the fields to be constant. The relevant

Feynman integral is

$$\frac{2M}{f}\,\mathrm{tr}\left(\Pi\, a_{\mu_1}\, a_{\mu_2}\, a_{\mu_3}\, a_{\mu_4}\right)$$

$$\int \mathrm{tr}\left(\gamma^{\mu_4}\gamma_5\,\frac{\not k+M}{k^2-M^2}\,\gamma_5\,\frac{\not k+M}{k^2-M^2}\,\gamma^{\mu_1}\gamma_5\,\frac{\not k+M}{k^2-M^2}\,\gamma^{\mu_2}\gamma_5\,\frac{\not k+M}{k^2-M^2}\,\gamma^{\mu_3}\gamma_5\,\frac{\not k+M}{k^2-M^2}\right)\frac{d^4k}{(2\pi)^4} \tag{6a.3.46}$$

The integrand is

$$\begin{aligned} &-\mathrm{tr}\left(\gamma^{\mu_4}\,\frac{1}{k^2-M^2}\,\gamma^{\mu_1}\gamma_5\,\frac{\not k+M}{k^2-M^2}\,\gamma^{\mu_2}\gamma_5\,\frac{\not k+M}{k^2-M^2}\,\gamma^{\mu_3}\gamma_5\,\frac{\not k+M}{k^2-M^2}\right)\\ &= -\mathrm{tr}\left(\gamma^{\mu_4}\,\frac{1}{k^2-M^2}\,\gamma^{\mu_1}\,\frac{\not k-M}{k^2-M^2}\,\gamma^{\mu_2}\,\frac{\not k+M}{k^2-M^2}\,\gamma^{\mu_3}\gamma_5\,\frac{\not k+M}{k^2-M^2}\right)\end{aligned} \tag{6a.3.47}$$

Averaging over directions of k gives

$$\begin{aligned} &\mathrm{tr}\left(\gamma^{\mu_4}\,\frac{1}{k^2-M^2}\,\gamma^{\mu_1}\,\frac{M}{k^2-M^2}\,\gamma^{\mu_2}\,\frac{M}{k^2-M^2}\,\gamma^{\mu_3}\gamma_5\,\frac{M}{k^2-M^2}\right)\\ &-\frac{1}{4}\,k^2\mathrm{tr}\left(\gamma^{\mu_4}\,\frac{1}{k^2-M^2}\,\gamma^{\mu_1}\,\frac{\gamma_\mu}{k^2-M^2}\,\gamma^{\mu_2}\,\frac{\gamma^\mu}{k^2-M^2}\,\gamma^{\mu_3}\gamma_5\,\frac{M}{k^2-M^2}\right)\\ &-\frac{1}{4}\,k^2\mathrm{tr}\left(\gamma^{\mu_4}\,\frac{1}{k^2-M^2}\,\gamma^{\mu_1}\,\frac{\gamma_\mu}{k^2-M^2}\,\gamma^{\mu_2}\,\frac{M}{k^2-M^2}\,\gamma^{\mu_3}\gamma_5\,\frac{\gamma^\mu}{k^2-M^2}\right)\\ &+\frac{1}{4}\,k^2\mathrm{tr}\left(\gamma^{\mu_4}\,\frac{1}{k^2-M^2}\,\gamma^{\mu_1}\,\frac{M}{k^2-M^2}\,\gamma^{\mu_2}\,\frac{\gamma_\mu}{k^2-M^2}\,\gamma^{\mu_3}\gamma_5\,\frac{\gamma^\mu}{k^2-M^2}\right)\end{aligned} \tag{6a.3.48}$$

The third term vanishes and the second and fourth and equal, so this is

$$\begin{aligned} &\mathrm{tr}\left(\gamma^{\mu_4}\,\frac{1}{k^2-M^2}\,\gamma^{\mu_1}\,\frac{M}{k^2-M^2}\,\gamma^{\mu_2}\,\frac{M}{k^2-M^2}\,\gamma^{\mu_3}\gamma_5\,\frac{M}{k^2-M^2}\right)\\ &+k^2\mathrm{tr}\left(\gamma^{\mu_4}\,\frac{1}{k^2-M^2}\,\gamma^{\mu_1}\,\frac{1}{k^2-M^2}\,\gamma^{\mu_2}\,\frac{1}{k^2-M^2}\,\gamma^{\mu_3}\gamma_5\,\frac{M}{k^2-M^2}\right)\end{aligned} \tag{6a.3.49}$$

which gives

$$-\,4i\,\epsilon^{\mu_1\mu_2\mu_3\mu_4}\left\{\frac{M^3}{(k^2-M^2)^4}+\frac{M\,k^2}{(k^2-M^2)^4}\right\} \tag{6a.3.50}$$

Now again using (6a.3.17) and (6a.3.35) and putting in the color factor, we get

$$-\,\frac{1}{4\pi^2 f}\,\mathrm{tr}\left(\Pi\, a_{\mu_1}\, a_{\mu_2}\, a_{\mu_3}\, a_{\mu_4}\right)\epsilon^{\mu_1\mu_2\mu_3\mu_4} \tag{6a.3.51}$$

Problems

6a-1. Repeat the Steinberger calculation of $\pi^0 \rightarrow \gamma\gamma$.

6a-2. Do the analog of the Steinberger calculation for the "anomalous" interaction term of (6a.1.30). That is, compute the contribution to such a term from a loop of quarks coupled (non-derivatively) to the Goldstone bosons, as in the q' quarks of (6a.2.4). Show that it reproduces the Wess-Zumino result. **Hint:** If you think carefully about all the simplifications that are possible because of the factor of $\epsilon_{\mu\nu\alpha\beta}$, the Feynman graph is extremely simple. Think, rather than working too hard.

6a-3. Use (6a.1.28) to calculate the decay rates for $\pi^0 \rightarrow 2\gamma$ and $\eta \rightarrow 2\gamma$. Compare your results with the data in the particle data book. Can you suggest any reason why the π^0 rate should be more reliably given by (6a.1.28) than the η rate?

7 — The Parton Model

7.1 Mode Counting

The basic idea of the parton model can be easily understood in the toy QCD model of Section 3.2. In the imaginary world with weak QCD, it is easy to calculate, for example, all weak decay rates. Just calculate in the quark theory using perturbation theory. This gives a perfectly adequate description of what happens during the short time that the particles are actually decaying. Much later (very much later in the imaginary world) when the decay products separate to distances at which confinement becomes important, complicated things happen (glueball emission, in the imaginary world), but these don't effect the total rate. The same ideas can be applied to our world.

Many processes involving quarks in the real world can be characterized by two separate time scales: a short time during which the quarks and gluons interact relatively weakly because of asymptotic freedom, and a longer time scale at which confinement effects become important. When these two scales can be cleanly separated, we can give the process a parton-model interpretation.

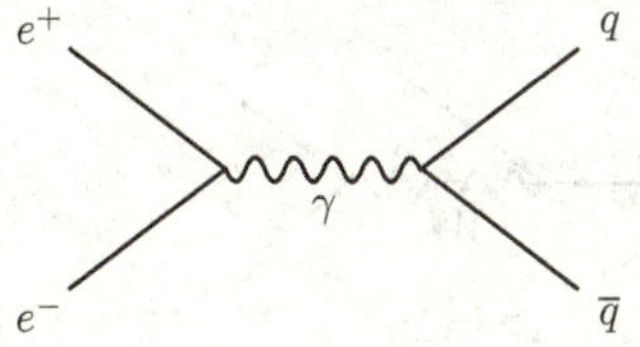

Figure 7-1:

The simplest parton-model process is the e^+e^- total cross section into hadrons. The relevant Feynman diagram is Figure 7-1 , where q is any quark that is light enough to be produced. If we compare this with $e^+e^- \to \mu^+\mu^-$ (Figure 2-4), we see that if the center of mass energy is large compared to the μ and quark masses, the only difference between the Feynman diagrams is the charge and color of the quark. The charge Q produces an extra factor of Q^2 in the quark cross section. The color produces an extra factor of 3, just because we must sum over the three colors in the final state. Thus, the ratio of the quark to muon total cross sections in lowest order is

$$R = \sum_{\text{quarks}} 3Q^2, \tag{7.1.1}$$

where the sum runs over all quarks with mass less than half the center of mass energy (that is the electron or positron beam energy , of course) , $E/2$.

This prediction for the total cross section can be improved by including QCD corrections, such as the single-gluon emission diagram (Figure 7-2). This is rather easy. When the appropriate

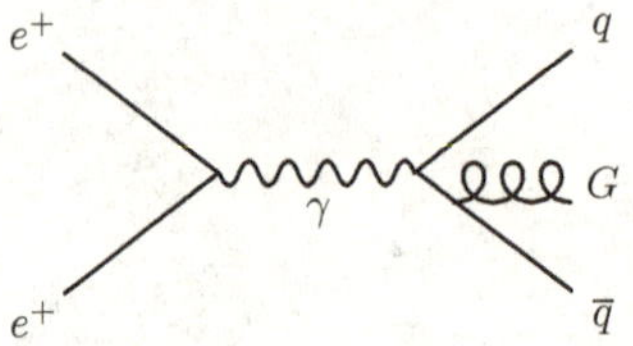

Figure 7-2:

virtual gluon diagrams are included, the result is

$$R = \sum sQ^2 \left[1 + \alpha_s/\pi + 0\left((\alpha_s/\pi)^2\right)\right] . \tag{7.1.2}$$

The higher-order terms are harder, but the order $(\alpha_s/\pi)^2$ terms have been calculated. The coupling in (7.1.2) should be evaluated at the center of mass energy, thus the QCD corrections go to zero as the energy gets Large. There are also corrections to (7.1.2) that cannot be calculated in perturbation theory in α_s. But these all involve the confinement time scale, and they fall off like powers of $(\Lambda/E)^2$ Thus, at large E, the simple parton prediction (7.1.1) should be accurate.

Indeed, the data bears out this simple picture. Except at low energies or near the threshold for production of new quark states, the cross section is roughly flat and roughly given by (7.1.1). In fact, the observed cross sections are always a bit Larger than (7.1.1), as we would expect from the form of the QCD corrections in (7.1.2). Near thresholds, complicated things happen, most conspicuously production of quark-antiquark resonances like ρ^0, ω, ϕ, J/ψ, and Υ all of which give rise to large, relatively narrow peaks in R.

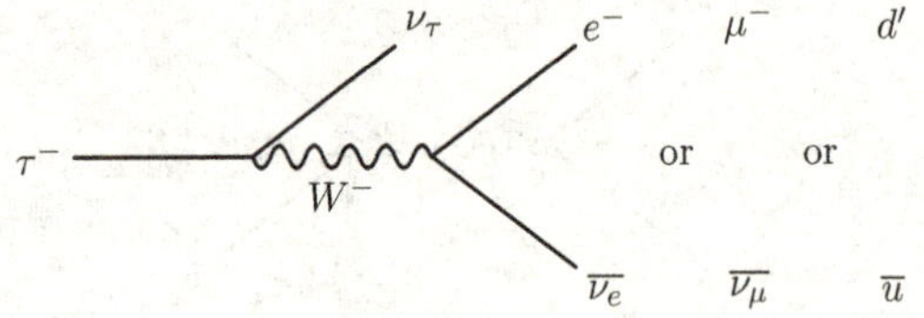

Figure 7-3:

As discussed in Section 2.9, weak-interaction effects have already been seen in $e^+e^- \to \mu^+\mu^-$. It is somewhat more difficult to see (and interpret) the analogous effects in $e^+e^- \to q\overline{q}$. However, there are processes in which exactly the same parton-model ideas can be employed to predict weak effects, for example, decay of a heavy lepton such as the τ^- . In this process, the τ^- decays into a ν_τ, and a virtual W^-. The subsequent splitting of the virtual W^- is precisely analogous to the decay of the virtual photon shown in Figure 7-1. Here the appropriate Feynman diagram is shown in Figure 7-3, where d' is the combination of d and s (the b is too heavy to be produced) that appears in the $SU(2)$ doublet with the u quark. To first approximation, we can take $d' = d$. Now the same kind of mode counting applies as in e^+e^- annihilation, but it is even easier here because all the couplings are the same. Thus, we expect branching ratios of 20% each into electrons and muons and 60% into primarily nonstrange hadrons. QCD corrections will modify this picture slightly by enhancing the hadronic modes as in (7.1.2). Thus, the leptonic modes should be somewhat smaller than 20% but, of course, still equal.

This is exactly what is observed. The leptonic branching ratios are about 18% each, and the hadronic modes consist primarily of nonstrange hadrons.

There is even some evidence that the d' linear combinations of d and s is the same as that measured in decays of strange particles. There we found, with

$$d' = \cos\theta_1 d + \sin\theta_1 \cos\theta_3 s, \tag{7.1.3}$$

that

$$\cos\theta_1 = 0.97, \quad \sin\theta_1 \, , \cos\theta_3 = 0.22, \tag{7.1.4}$$

so that the ratio of us to rd in the final state of hadronic τ^- decay should be

$$\frac{\sin^2\theta_1 \cos^2\theta_s}{\cos^2\theta_1} = 0.05. \tag{7.1.5}$$

The decay

$$\tau^- \to \nu_\tau + K^{*-}(892) \tag{7.1.6}$$

has been observed. The $K^{*-}(892)$ is a $\overline{u}s$ bound-state vector meson that is the analog of the $\overline{u}d$ bound-state $\rho^-(770)$. Thus, we would expect the decay (7.1.6) to have a branching ratio smaller by a factor of 20 than the decay

$$\tau^- \to \nu_\tau + \rho^-(770). \tag{7.1.7}$$

Again, this is just what is seen. (7.1.6) has branching ratio $(1.45 \pm 0.18)\%$ while (7.1.7) has branching ratio about 25% (the $\pi^-\pi^0\nu_\tau$ mode, with branching ratio $(25.2 \pm 0.4)\%$ is essentially all $\rho^-\nu_\tau$). Apparently, both the parton model and the $SU(2) \times U(1)$ model work very well for heavy lepton decay.

7.2 Heavy Quark Decay

One might imagine that the decay of a hadron containing a heavy quark such as a t, b, or c might be described in the same way by a diagram such as Figure 7-4. In this diagram the $\overline{q}_{\text{light}}$ plays no role. Even higher-order QCD corrections, in which gluons are exchanged between $\overline{q}_{\text{light}}$ and everything else, do not depend on the identity of $\overline{q}_{\text{light}}$. Thus, the $\overline{q}_{\text{light}}$ is just a spectator. We will call processes like that in Figure 7-4 "spectator processes".

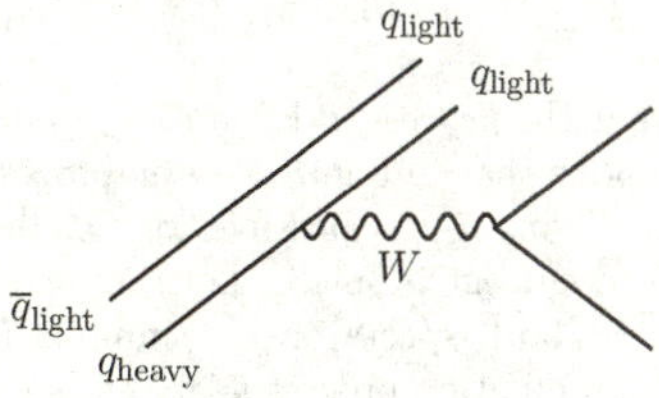

Figure 7-4:

It is clear that lots of other things can happen in the decay of a hadron containing a heavy quark. In particular, the presence of the other quark or antiquark in the initial state can influence the gross structure of the decay. The most extreme example of this is a purely leptonic decay . We would expect, for example, that the F^+ meson will decay occasionally into $\mu^+\,\nu$ through a diagram (Figure 7-5) in which the c and $\overline{s}$ in the F^+ annihilate into a virtual W^+.

However, this "annihilation diagram" will be suppressed because of angular-momentum conservation, just like the decay $\pi^+ \to e^+\nu$. However, there is a class of generalized annihilation processes that are not as suppressed as what we find in Figure 7-5, for example, the processes in Figures 7-6 and 7-7.

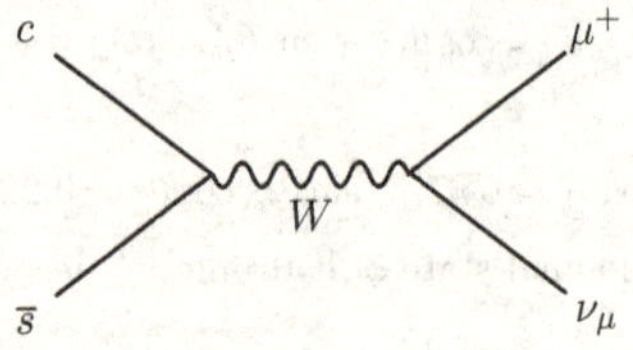

Figure 7-5:

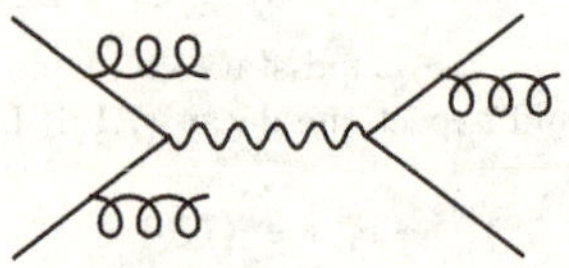

Figure 7-6:

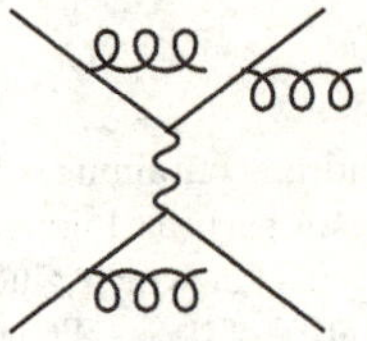

Figure 7-7:

Indeed, there is evidence that the process in Figure 7-7 is important in charmed particle decay. To understand the evidence, notice that in Figure 7-7 the process is possible for D^0 decay but not for D^+ decay. The process in Figure 7-6 is possible for D^+ decay, but it is suppressed because the angle associated with the $c\overline{d}$ current is small. Thus, we expect only spectator interactions to be important for D^+ decay, but both spectator and annihilation to be important in D^0 decay. Experiment suggests that the annihilation process is about as important as the spectator process. The evidence for this comes from a measurement of the semileptonic branching ratios of the D^0 and D^+ and from measurements that suggest that the D^+ lifetime is longer than that of the D^0.

The semileptonic branching ratios are measured to be

$$\begin{aligned} B(D^+ \to e^+ \quad +\text{anything}) \quad (17.2 \pm 1.9)\% \\ B(D^0 \to e^+ \quad +\text{anything}) \quad (7.7 \pm 1.2)\%. \end{aligned} \tag{7.2.1}$$

The D^+ decay is about what we would expect on the basis of the spectator process. This is reasonable because the annihilation contribution is suppressed. But the D^0 decay is quite different. Presumably, the spectator decay contributes to $D^0 \to e^+ +$ anything at the same rate as $D^+ \to e^+ +$ anything anything. The branching ratio in D^0 decay must be smaller because the annihilation process is making a large contribution to the nonleptonic modes. This, of course, must make the D^0 decay rate larger.

Frankly, the size of the annihilation effect in (7.2.1) is something of an embarrassment. No one really knows how to calculate it, but theoretical estimates before the measurement put the annihilation contribution as smaller than or perhaps at most equal to the spectator contribution. But that means that the ratio of the D^+ to D^0 branching ratios in (7.2.1) should have been no more than a factor of two. We have now gotten used to the experimental ratio, but I still believe that we don't know how to calculate it.

The data from direct measurements of the D^+ and D^0 lifetimes are now pretty good. Because of improvement in detector technology, we can now measure these very short lifetimes. On the basis of the spectator model and mode counting, we would expect a lifetime of order

$$\tau_\mu \left(\frac{m_\mu}{m_c}\right)^5 \cdot \frac{1}{5} \simeq 8 \times 10^{-13} \text{ sec}, \tag{7.2.2}$$

where m_c is the charmed quark mass, about 1.5 GeV. Initially, the experiments showed a D^0 lifetime much shorter than this, consistent with the semileptonic decay data. But more recent experiments tend to support the view that the D^0 lifetime is about a factor of two shorter than that of the D^+. The most recent numbers are:

$$\begin{aligned} \tau_{D^+} &= (1.057 \pm 0.015) \times 10^{-12} \text{ sec} \\ \tau_{D^0} &= (.415 \pm 0.004) \times 10^{-12} \text{ sec}. \end{aligned} \tag{7.2.3}$$

Note that τ_{D^+} is roughly consistent with the spectator-model prediction, (7.2.2).

For the b quark (and heavier quarks), we expect the spectator model to work well Fragmentary data on b-quark decay suggests that the b is rather long-lived, with a lifetime of order 10^{-12} sec. This would imply small values for s_2 and s_3 in the KM matrix.

7.3 Deep Inelastic Lepton-Hadron Scattering

One reason for the tension that you may have noticed between theorists and experimentalists is a very natural difference between their short-term goals. Experimenters want to do experiments that are doable while theorists want to interpret experiments that are interpretable. This is a real problem in experiments with hadrons because the quantities that are easy to measure are typically related to the long-distance structure of the theory that is too hard for the theorists to deal with.

Deep inelastic lepton-hadron scattering experiments are a class of experiments about which experimenters and theorists can both be enthusiastic. The idea is to scatter a high-energy lepton beam off a hadron target and measure the four momentum of the scattered lepton but not the details of the hadronic final state. A typical process is electron-proton scattering, first studied extensively with the high-energy electron beam from the two-mile-long linear accelerator at SLAC in Stanford.

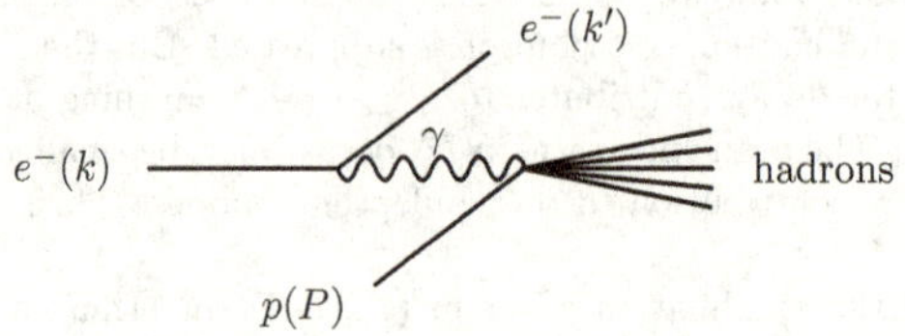

Figure 7-8:

The dominant process is the electromagnetic scattering described by the diagram in Figure 7-8. The momentum P' of the hadronic final state is determined by momentum conservation, $P' = P + k - k'$, but no other information about the state is used. This is just as well because it is usually very complicated, involving lots of pions. The virtual photon carries momentum q, which is spacelike, so $-q^2$ is positive. The product Pq is the proton mass m times the virtual photon energy, so it is positive (the photon energy is sometimes called ν). It is conventional and useful to define the dimensionless variables:

$$x_H = -\frac{q^2}{2Pq}, \quad y = \frac{Pq}{Pk}. \tag{7.3.1}$$

Both of these are bounded between zero and one. If E and E' are the energies of the initial and final electron, y is

$$y = \frac{E - E'}{E}, \tag{7.3.2}$$

the fractional energy loss of the lepton. When $y = 0$, no energy is transferred to the hadronic system (which is impossible kinematically, but one can get very close to $y = 0$ at high lepton energies). When $y = 1$, the initial lepton energy is almost all transferred.

To see that x_H must be less than 1 , note that $P' = P + q$ and

$$P'^2 = m^2 + 2Pq + q^2 \geq m^2, \tag{7.3.3}$$

since there is always at least one proton in the final state. This immediately gives $x_H \leq 1$.

We will study the differential cross sections $d^2\sigma/dx_H dy$. To interpret the results, we need a theoretical model that describes the process in QCD. The key idea, due to Feynman, is that there are two distinct time scales characterizing this process. The virtual photon that is far off its mass shell scatters off a quark, antiquark, or gluon in a short time of order $1/\sqrt{-q2}$. The hadronic time scale Λ enters in two ways. The quarks have been bound in the hadron for a long time, and they may be off their mass shell by a momentum of order Λ. Also, after scattering, the quarks recombine into hadrons in a time of order $1/\Lambda$.

When Feynman invented the idea, he wasn't sure what kind of particles inside the proton were scattering. He just called them partons, the parts of the proton. Gell-Mann, meanwhile, knew all about quarks, but it took almost ten years for the two of them to get together. (J. D. Bjorken was one of the first to put these ideas together.)

At any rate, the presence of these two scales suggests that we view the scattering as a two-step process in which first the short-distance scattering occurs, then later the scattered parton and the debris of the shattered proton recombine into hadrons. As in e^+e^-, we assume that the cross section is determined by the short-distance scattering. To calculate it, we need to know two things: the probability of finding a given parton in the proton carrying a given momentum, and the probability of the parton scattering from the virtual photon.

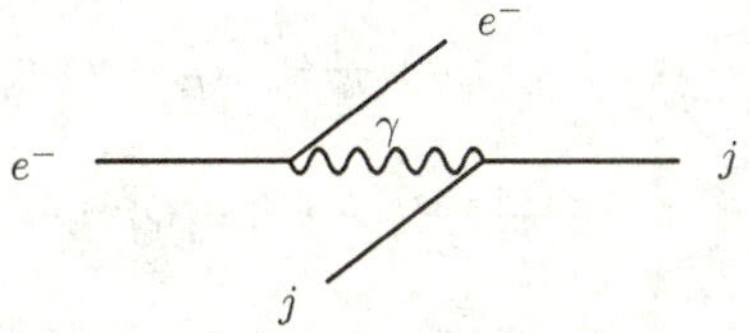

Figure 7-9:

We will analyze the process in the center of mass frame of the proton and the virtual photon because it is only through the virtual photon that momentum is transferred to the hadronic system. At high Pq, the proton is going very fast in this frame, and the partons have momenta that are parallel to the proton's momentum P up to terms of order Λ. We will ignore these (because they change the cross section only by $\sim \Lambda^2/-q^2$ and write the momentum of the a parton as

$$p^\mu = \xi P^\mu, \tag{7.3.4}$$

where $0 \leq \xi \leq 1$. Then we can describe the distribution of parton in the proton as a function of the single variable ξ. We write

$$f_j(\xi)\, d\xi \tag{7.3.5}$$

is the probability of finding a parton of type j (j runs over each type of quark and antiquark and gluon) with momentum between ξP and $(\xi + d\xi)P$. This defines the distribution function $f_j(\xi)$. These cannot be calculated with our present theoretical tools. They depend on the structure of the proton, which we do not understand in terms of QCD perturbation theory.

We can, however, calculate the probability for parton scattering. We do this just by calculating the cross section $d\sigma_j/dx\,dy$, where x and y are defined in terms of the parton momentum $p^\mu = \xi P^\mu$,

$$\begin{aligned} x &= -\tfrac{q^2}{2pq} = x_H/\xi \\ y &= \tfrac{pq}{pk} = \tfrac{Pq}{Pk}, \end{aligned} \tag{7.3.6}$$

and the j-type parton is the target. The lowest-order diagram contributing to this process is that in Figure 7-9, where j is a quark or antiquark. I will not calculate this in detail, but it is fairly easy to figure out its general form. $d\sigma_j$ is clearly proportional to $e^4Q_j^2/q^4$, where Q_j is the quark or antiquark charge. The kinematics produces a factor of $\left[(1+(1-y)^2\right]$ for reasons that we will describe in detail later. There is also a δ function that expresses the fact that the final quark in Figure 7-9 is on its mass shell,

$$\delta(p'^{\,2}) = \delta(2pq + q^2) \propto \delta(1-x), \tag{7.3.7}$$

(where we have neglected the quark mass, though it wouldn't matter). Finally, to get the dimensions right, we need a factor of $k \cdot p = \xi k \cdot P = \xi m E$. Thus, putting the factors of π and 2 in, we get

$$\frac{d\sigma_1}{dx\, dy} = \frac{e^4 Q_j^2}{4q^4} \frac{\xi m E}{\pi} \left[1 + (1-y)^2\right] \delta(x-1). \tag{7.3.8}$$

We now can put this together just by combining the probabilities, using the fact that $x_H = \xi x$, to obtain (see Problem 7-1)

$$\begin{aligned} \frac{d\sigma_H}{dx_H\, dy} &= \textstyle\sum_j \int f_j(\xi) \cdot \frac{d\sigma_j}{dx\, dy}\, \delta(x_H - x\xi)\, d\xi\, dx \\ &= \frac{e^4 m E}{4\pi q^4} [1 + (1-y)^2] \textstyle\sum_j Q_j^2 x f_j(x). \end{aligned} \tag{7.3.9}$$

The striking thing about this result is that the dependence on q^2 is like that in electron scattering off a pointlike target. This is just what is observed.

Higher-order QCD effects change this simple prediction only slightly. The distribution functions actually are predicted to depend weakly on q^2. Something very much like the predicted q^2 dependence has actually been observed. This subject is very interesting but should be discussed in a course on strong interactions. Here we will work only to zeroth order in QCD and use (7.3.9).

7.4 Neutrino-Hadron Scattering

Precisely the same reasoning can be applied to neutrino-hadron or anti-neutrino-hadron scattering. The new elements here are $W^\pm$ or Z^0 exchange rather than photon exchange and the parity-violating nature of the couplings.

We will first consider the charged-current process ν_μ (or $\overline{\nu}_\mu$) + hadron $\rightarrow \mu^-$ or μ^+) + anything. This is easier to see experimentally than the neutral-current processes we will discuss later. The parton processes differ depending on whether the beam is ν_μ, or $\overline{\nu}_m u$. For ν_μ, the important processes are

$$\begin{aligned} \nu_\mu + d &\rightarrow \mu^- + u \\ \nu_\mu + \overline{u} &\rightarrow \mu^- + \overline{d}, \end{aligned} \tag{7.4.1}$$

while for $\overline{\nu}_\mu$, they are

$$\begin{aligned} \overline{\nu}_\mu + u &\rightarrow \mu^+ + d \\ \overline{\nu}_\mu + \overline{d} &\rightarrow \mu^+ \overline{u}. \end{aligned} \tag{7.4.2}$$

The cross sections for these parton processes are similar to those for e^--quark scattering, but they differ in two ways. The $W^\pm$ propagator gives a G_F^2 rather than e^4/q^4, at least at $-q^2$ small compared to M_W^2. And the $(1+\gamma_5)$'s in the weak coupling affect the y distribution. The rule is the following: when particles of the same handedness scatter, as $\nu_\mu + d$ (both left-handed) or

$\overline{\nu}_\mu + bard$ (both right-handed), the cross section is independent of y; when particles of the opposite handedness scatter, as $\nu_\mu + \overline{u}$ or $\overline{\nu}_\mu + u$, the cross section vanishes at $y = 1$ like $(1-y)^2$.

Putting all this together, we find for the hadronic cross sections

$$\frac{d\sigma^\nu_H}{dx_H\,dy} = \frac{2mEG_F^2}{\pi} \cdot \left\{xd(x) + (1-y)^2 x\overline{u}(x)\right\} \tag{7.4.3}$$

$$\frac{d\sigma^{\overline{\nu}}_H}{dx_H\,dy} = \frac{2mEG_F^2}{\pi} \cdot \left\{(1-y)^2 xu(x) + x\overline{d}(x)\right\}. \tag{7.4.4}$$

We have ignored various small effects, such as the distribution functions of strange and charmed quarks and the strangeness-changing and charm-changing parts of the u and d currents, but (7.4.3)-(7.3.4) is a good first approximation, and the other effects can be included by the same techniques.

The distribution functions depend on the target. Things are particularly simple for targets that consist of equal numbers of protons and neutrons. Then isospin symmetry implies

$$u = d \equiv q, \quad \overline{u} = \overline{d} \equiv \overline{q}. \tag{7.4.5}$$

We expect q to be much larger than $\overline{q}$ and perhaps also to have a different shape. What is seen is that $\overline{q}(x)$ is smaller than $q(x)$ and vanishes faster as x goes to one.

Notice that the total cross sections are (with $f_q = \int dx\; 2xq(x)$, and so on)

$$\begin{aligned} \sigma^\nu &= \tfrac{mE}{\pi} G_F^2 [f_q + f_{\overline{q}}/3] \\ \sigma^{\overline{\nu}} &= \tfrac{mE}{\pi} G_F^2 [f_q/3 + f_{\overline{q}}], \end{aligned} \tag{7.4.6}$$

where f_q (and $f_{\overline{q}}$) are the fraction of the proton and neutron momentum carried by quarks (and antiquarks). This can be directly compared with the e^- cross section n the same approximation of ignoring heavy quarks:

$$\sigma^{e^-} = \frac{e^4 mE}{4\pi q^4}\,\frac{4}{3} \cdot \frac{5}{18}\,(f_q + f_{\overline{q}}). \tag{7.4.7}$$

Thus, we find the relation (which works)

$$\sigma^{e^-} = \frac{5}{72}\,\frac{e^4}{q^4 G_F^2}\,(\sigma^\nu + \sigma^{\overline{\nu}}). \tag{7.4.8}$$

From the observed cross sections, we can also extract the normalized values of f_q and $f_{\overline{q}}$. The results are (very roughly, since we have not included heavy quarks or QCD corrections)

$$f_q \approx 0.5, \quad f_{\overline{q}} \approx 0.1 \tag{7.4.9}$$

The amusing thing here is that the quarks and antiquarks carry only about half of the nucleon's momentum. Most of the rest is carried by gluons. Heavy quarks and antiquarks carry a small fraction.

The fact that the parton-lepton scattering vanishes as $y \to 1$ when the lepton and parton have opposite handedness can be seen easily by a helicity argument. $y = 1$ corresponds to a lepton that loses all its energy in the lab frame. In the center of mass frame, that means the particles are scattered backwards. It is then easy to see that the scattering is allowed by angular-momentum conservation about the beam axis if the parton and lepton have the same handedness, but not if they have opposite handedness. This is the physical reason for the factors of $(1-y)^2$ in (7.3.8)-(7.3.9) and (7.4.3)-(7.4.4).

7.5 Neutral Currents

In much the same way, we can study neutral-current processes

$$\begin{aligned} \nu_\mu + P \text{ or } N \quad &\to \nu_\mu + \text{anything} \\ \nu_\mu + P \text{ or } N \quad &\to \overline{\nu}_\mu + \text{anything.} \end{aligned} \tag{7.5.1}$$

The cross sections are similar to the charged current but involve the coupling of the Z^0 to the quarks, $T-3 \to \sin^2\theta Q$. This time both left- and right-handed quarks scatter, but T_3 is nonzero only for the left-handed quarks. Thus, in the same approximation we used for (7.4.3)-(7.4.4), we have

$$\begin{aligned} \frac{d\tilde{\sigma}_H}{dx_H dy} = \frac{2mEG_F^2}{\pi} \cdot \{ &(\epsilon_L(u) + (1-y)^2\epsilon_R(u))\, x\, u(x) \\ &+ ((1-y)^2\epsilon_L(u) + \epsilon_R(u))\, x\overline{u}(x) \\ &+ (\epsilon_L(d) + (1-y)^2\epsilon_R(d))\, x\, d(x) \\ &+ ((1-y)^2\epsilon_L(d) + \epsilon_R(d))\, x\, \overline{d}(x)\}, \end{aligned} \tag{7.5.2}$$

where $\epsilon_{L,R}(q)$ is $(T_3 - \sin^2\theta\, Q)^2$ for the appropriate quark state.

$$\begin{aligned} \epsilon_L(u) &= \left(\tfrac{1}{2} - \tfrac{2}{3}\sin^2\theta\right)^2 \\ \epsilon_L(d) &= \left(-\tfrac{1}{2} + \tfrac{1}{3}\sin^2\theta\right)^2 \\ \epsilon_R(u) &= \left(-\tfrac{2}{3}\sin^2\theta\right)^2 \\ \epsilon_R(d) &= \left(\tfrac{1}{3}\sin^2\theta\right)^2 . \end{aligned} \tag{7.5.3}$$

The antineutrino cross section $d\tilde{\sigma}^{\overline{\nu}}/dx_H dy$ can be obtained from (7.5.2) by interchange of the quark and antiquark distributions.

Things simplify if we look at the total cross-sections off matter targets ($u = d$). Then

$$\tilde{\sigma}^\nu = \frac{mEG_F^2}{\pi}\left\{(\epsilon_L(u) + \epsilon_L(d))\left[f_q + \frac{1}{3}f_{\overline{q}}\right] + (\epsilon_R(u) + \epsilon_R(d))\left[\frac{1}{3}f_q + f_{\overline{q}}\right]\right\} \tag{7.5.4}$$

and again $\tilde{\sigma}^{\overline{\nu}}$ is obtained by interchanging q and $\overline{q}$. The ratios of the neutral to charged-current cross sections are useful because some experimental uncertainties drop out. The predictions are

$$R_\nu = \tilde{\sigma}^\nu/\sigma^\nu = (\epsilon_L(u) + \epsilon_L(d)) + x^{-1}(\epsilon_R(u) + \epsilon_R(d))$$

$$R_{\overline{\nu}} = \tilde{\sigma}^{\overline{\nu}}/\sigma^\nu = (\epsilon_L(u) + \epsilon_L(d)) + x(\epsilon_R(u) + \epsilon_R(d)). \tag{7.5.5}$$

where

$$x = \sigma^\nu/\sigma^{\overline{\nu}} = \frac{f_q + \frac{1}{3}f_{\overline{q}}}{\frac{1}{3}f_q + f_{\overline{q}}}. \tag{7.5.6}$$

The measured values of R_ν and $R_{\overline{\nu}}$ agree with the predictions of the standard $SU(2) \times U(1)$ model and the quark-parton model for $\sin^2\theta \simeq 0.23$.

7.6 The SLAC Experiment

The linear accelerator at SLAC, which was responsible for most of the early experimental work on $e^- P$ scattering that gave rise to the parton model, was also the site of the crucial experiment that convinced us of the correctness of the $SU(2) \times U(1)$ model. The experiment measured the difference between the scattering cross sections for left-handed and right-handed electrons on protons. This is a parity-violating effect, obviously, so the leading electromagnetic contribution to the cross section cancels out in the difference. What is left is a weak-electromagnetic interference that can be calculated using standard parton-model techniques. This provides very different information from neutrino scattering, since it is linear in the quark-Z^0 coupling and probes the structure of the e^- neutral current. The results are in agreement with the $SU(2) \times U(1)$ predictions with $\sin^2\theta \simeq 0.23$ (see C. Y. Prescott *et al.*, *Physics Letters* **84B**, 524-528, 1979) (see Problem 7-2).

Problems

7-1. Derive (7.3.9) and (7.4.3)-(7.4.4).

7-2. By including Z^0 exchange in the $e^- p$-scattering cross section, calculate the parity-violating $\gamma - Z^0$ interference contribution to the difference in differential cross sections for left- and right-handed electrons,

$$\frac{\partial\sigma_L}{\partial dx_H dy} - \frac{\partial\sigma_R}{\partial dx_H dy}$$

8 — Standard Model Precision Tests

Since the original version of this text was written, experimental tests of the standard model have progressed to the point that there is overwhelming evidence for the basic picture of a spontaneously broken $SU(2) \times U(1)$ gauge theory. The Z and W have been studied extensively, and their static properties (masses, partial widths into various decay channels, and so on) have been measured. There are still details that we are unsure about (such as the precise couplings of the W to quarks), but in general, the experimental results agree with the theoretical predictions of the simplest version of the standard model not just in the tree approximation but beyond, to the level of the radiative corrections. Nevertheless, we will argue that it is still unclear what the physics is that spontaneously breaks the $SU(2) \times U(1)$ gauge symmetry, for two reasons:

1. The largest radiative corrections are independent of the details of electroweak symmetry breaking, depending only on the particle content of the theory below the symmetry breaking scale; and
2. The leading low-energy effects of the physics of electroweak symmetry breaking are relatively small, and tightly constrained by the symmetries of the model. They can be gathered into the values of a few effective interactions.

We will only know for sure what breaks the electroweak symmetry when we can see the scattering of Ws and Z at very high energies, of the order of 1 TeV. This should be possible at the LHC - the Large Hadron Collider, expected to operate at CERN in the first decade of the next millennium.

It this chapter, we will review some of the issues involved in the calculation of radiative corrections, and show in what sense the largest corrections are simple, and independent of the details of symmetry breaking. We will also briefly discuss two alternatives to the simplest version of electroweak symmetry breaking — technicolor and a composite Higgs boson. We will not attempt to give a complete account of modern calculations of radiative corrections. This has become a large industry, and the details are beyond the scope of this course.

8.1 Choosing a Gauge

In a theory like $SU(2) \times U(1)$ with spontaneously broken gauge symmetry, the unitary gauge that we introduced in Section 2.7 is not very convenient for explicit loop calculations. This formalism has the advantage that the only fields that appear correspond to physical particles. But the resulting Lagrangian involves massive vector fields coupled to nonconserved currents. As we discussed in Section 2.3, this is a potentially dangerous situation. The renormalizability of the theory is not obvious. In this section, we will discuss a class of gauges in which renormalizability is more explicit.

In chapter 1b, we discussed the necessity of choosing a gauge and the Fadeev-Popov recipe for constructing a gauge fixing term.

The Fadeev-Popov "derivation" discussed in chapter 1b was very formal. The functional integral we started with was not well defined, and it is certainly not obvious that the result of interchanging order of functional integrations and canceling infinite quantities has anything to do with the gauge theory we started with. Physicists were impressed with the derivation because it incorporated all of the techniques that had been developed (unsystematically) over the years to deal with QED, Yang-Mills theories, and gravity. It was not taken for granted that it would be adequate to properly define spontaneously broken gauge theories. But at least for the gauge–fixing terms we will describe below, it does work. It yields theories that are unitary and renormalizable in perturbation theory.

We will now discuss some explicit gauge–fixing terms. The unitary gauge that we have already discussed in chapter 2 corresponds to a choice of the Fadeev-Popov function

$$f(A^\mu,\, \phi) = \prod_x \prod_{\substack{\text{broken}\\ \text{generators}}} \delta\left(\langle\phi^T\rangle g_a T_a \phi\right). \tag{8.1.1}$$

In this case Δ depends only on the surviving Higgs fields.

A gauge that is simple and convenient for some purposes is Landau gauge, which corresponds to

$$f(A^\mu,\, \phi) = \prod_x \prod_a \delta\,(\partial_\mu A^\mu_a)\,. \tag{8.1.2}$$

To calculate $\Delta(A^\mu)$, note that because of the appearance of the δ function (8.1.2) in the integral (1b.2.12), we need to know $\Delta(A^\mu)$ only for A^μ satisfying $\partial_\mu A^\mu_a = 0$. Thus, it is enough to consider infinitesimal transformations of the form

$$\Omega = 1 + i\omega_a T_a + \cdots \tag{8.1.3}$$

under which

$$A^\mu_{\Omega a} = A^\mu_a - f_{abc}\omega_b A^\mu_c - \frac{1}{g}\partial^\mu\omega_a \tag{8.1.4}$$

and

$$[d\Omega] = \prod_{a,\, x} [d\omega_a(x)]. \tag{8.1.5}$$

Then

$$f(A^\mu_\Omega,\, \phi_\Omega)[d\Omega] = \prod_{a,\, x}[d\omega_a(x)] \prod_{b,\, y} \delta\left(\partial_\mu A^\mu_{\Omega b}(g)\right). \tag{8.1.6}$$

If we insert (8.1.4) into (8.1.6) and ignore terms proportional to $\partial_\mu A^\mu_a$, we can write

$$\Delta(A^\mu)^{-1} = \int \prod_{b,\, y} \delta\left(\int M_{bc}(y,\, z)\omega_c(z) d^4z\right) \prod_{a,\, x}[d\omega_a(x)]\,, \tag{8.1.7}$$

where

$$M_{bc}(y,\, z) = -\frac{1}{g}\partial_\mu\left(\partial^\mu\delta_{bc} - g f_{bdc} A^\mu_d\right)\cdot\delta(y-z) = -\frac{1}{g}\partial_\mu D^\mu_{bc}(y-z). \tag{8.1.8}$$

Then

$$\Delta(A^\mu) = \det M. \tag{8.1.9}$$

We could calculate $\det M$ directly, but it is more fun to express it as a functional integral over anticommuting variables η_a and $\overline{\eta}_a$ in the adjoint representation of the gauge group:

$$\det M \propto \int \exp\{iS_{\text{Ghost}}\}\,[d\eta][d\overline{\eta}]\,, \tag{8.1.10}$$

where

$$S_{\text{Ghost}} = \int \overline{\eta}_a(x)\partial_\mu D^\mu_{ab}\eta_b(x)\, d^4x\,. \tag{8.1.11}$$

Thus, this term can simply be added to the action to produce the Fadeev-Popov determinant Δ. The $\overline{\eta}$ and η fields are the Fadeev-Popov ghost particles. They are massless scale fermions that (fortunately) do not appear in the physical states.

In the Landau gauge, the gauge–boson propagator is

$$\frac{-g^{\mu\nu} + k^\mu k^\nu/k^2}{k^2 - M^2}\,, \tag{8.1.12}$$

where M^2 is the gauge–boson mass–squared matrix. This gauge is extremely convenient for calculations in unbroken non-Abelian gauge theories, and it is also often useful in spontaneously broken gauge theories. In Landau gauge, the mixing terms between the gauge bosons and the Goldstone bosons, (2.6.21), are irrelevant because they are total divergences. Thus, while the Goldstone bosons are not eliminated, at least they do not complicate the gauge–boson propagators. However, for some applications, the appearance of the massless Goldstone bosons and the corresponding appearance of the pole at $k^2 = 0$ in the longitudinal part of the gauge–boson propagator is inconvenient. In such applications, it is often useful to use one of a set of gauges which interpolate continuously between Landau gauge and unitary gauge. These are the renormalizable ξ–gauges. We will discuss them in some detail, both because they are often useful and because they will give us some insight into the nature of the Higgs mechanism.

We start with an $f(A^\mu,\, \phi)$ of the form

$$\prod_x \prod_a \delta\left(\partial_\mu A^\mu_a + i\xi\langle\Phi^T\rangle g_a T_a \Phi - c_a\right), \tag{8.1.13}$$

where Φ is the unshifted Higgs field, c_a is an external field, and ξ is a positive constant. It may not be obvious why we have added the second and third terms in the δ functions at this point, but eventually we will use them to simplify the propagators. Calculating Δ by the techniques discussed above, we find (see Problem 8–1):

$$\Delta(A^\mu,\, \phi) \propto \tag{8.1.14}$$

$$\int \exp\left\{-i\int \overline{\eta}_a(x)\left[\partial_\mu D^\mu_{ab} + \xi\langle\Phi^T\rangle g_a T_a g_b T_b \Phi\right]\eta_b(x)\, d^4x\right\}[d\overline{\eta}][d\eta]\,.$$

Note that Δ is independent of c_a. Notice, also, that when we write Φ in terms of shifted fields, we are left in (8.1.14) with a mass term for the ghost fields. The mass–squared matrix is just

$$\mu^2_{ab} = \xi\langle\Phi^T\rangle g_a T_a g_b T_b\langle\Phi\rangle = \xi M^2_{ab}, \tag{8.1.15}$$

where M^2_{ab} is the gauge–boson mass–squared matrix.

Now consider the c_a dependence of a gauge–invariant Green's function, (1b.2.12). Clearly there isn't any, because the whole point of the Fadeev-Popov construction is that when Δ is properly chosen, it is irrelevant which gauge–fixing term we choose. Both W and G are independent of c_a, so long as g is gauge–invariant. Thus, in particular, we can multiply W and G by the quantity

$$\int \exp\left\{-\frac{1}{2\xi}\int c_a^2(x) d^4x\right\}[dc_a] \tag{8.1.16}$$

without changing anything. But then interchanging integration orders, we can do the integration of c_a and eliminate the δ functions. The result is a description of the theory in terms of a Lagrangian with the same ghost term (8.1.14), but with a gauge–fixing term of the form (in the Lagrangian)

$$-\frac{1}{2\xi}\left(\partial_\mu A_a^\mu + i\xi\langle\Phi^T\rangle g_a T_a \Phi'\right)^2 . \qquad (8.1.17)$$

Notice that we can write (8.1.17) in terms of the shifted field Φ' because of the antisymmetry of T_a:

$$\langle\Phi^T\rangle g_a T_a \langle\Phi\rangle = 0. \qquad (8.1.18)$$

Now finally, we can get to the physics. The cross term is (8.1.17) combines with the term (2.7.21) from the Higgs meson kinetic–energy term to form an irrelevant total–derivative term. Thus, we have succeeded in eliminating the mixing between gauge bosons and Goldstone bosons. The other terms in (8.1.17) fix the gauge and give mass to the Goldstone bosons. The mass matrix for the Goldstone boson fields is the same as that of the ghosts, μ^2_{ab} in (8.1.15). It sounds peculiar to speak of Goldstone–boson masses, but remember that the Goldstone bosons are gauge artifacts. With an appropriate choice of gauge, we can give them mass if we choose, or even eliminate them entirely as in unitary gauge. To underscore this point, I will usually refer to the Goldstone bosons in ξ–gauges as "unphysical Goldstone bosons".

The quadratic terms in the gauge fields are now

$$\begin{aligned} &\tfrac{1}{4}(\partial_\mu A_{a\nu} - \partial_\nu A_{a\mu})(\partial^\mu A_a^\nu - \partial^\nu A_a^\mu) \\ &-\tfrac{1}{2\xi}(\partial_\mu A_a^\mu)^2 + \tfrac{1}{2}M_{ab}^2 A_a^\mu A_{b\mu}. \end{aligned} \qquad (8.1.19)$$

This gives a propagator

$$\frac{[-g^{\mu\nu} + (1-\xi)k^\mu k^\nu/(k^2-\xi M^2)]}{[k^2-M^2]} \qquad (8.1.20)$$

or equivalently

$$= \frac{[-g^{\mu\nu} + k^\mu k^\nu/M^2]}{k^2-M^2} - \frac{k^\mu k^\nu/M^2}{k^2-\xi M^2}. \qquad (8.1.21)$$

(8.1.21) is extremely suggestive. The first term is simply the propagator for the massive vector fields. The second term modifies the longitudinal part of the gauge–boson propagator to make it well-behaved at large k^2. But this second term by itself looks as if it is related to propagation of derivatively coupled scalars with mass $\mu^2 = \xi M^2$, the same mass as the unphysical Goldstone bosons and the Fadeev-Popov ghosts. Indeed, the effects of these three gauge artifacts conspire to cancel in all physical processes so that unitarity is maintained in these gauges.

Practically speaking, the advantage of the ξ–gauges is that all gauge artifacts have nonzero masses. The simplest such gauge corresponds to $xi = 1$, for which the gauge–boson propagator is just

$$-\frac{g^{\mu\nu}}{k^2-M^2} \qquad (8.1.22)$$

and the unphysical Goldstone bosons and ghosts have the same masses as the corresponding gauge particles. This is the 't Hooft-Feynman gauge. For many purposes, it is the most convenient gauge for explicit calculations in a model like $SU(2)\times U(1)$. Some physicists prefer to keep ξ arbitrary, however, because the cancellation of all ξ dependence provides a useful check for complicated calculations.

Note that, as promised, these gauges interpolate between the Landau gauge ($\xi = 0$) and unitary gauge ($\xi = \infty$). The unitary–gauge limit is obtained by simply pushing all the gauge artifacts up to infinite mass.

Finally, note that the whole issue of gauge fixing is really not dependent on the assumption that a fundamental Higgs boson exists. It does depend on the Higgs mechanism, that describes how the Goldstone bosons associated with the spontaneous breaking of $SU(2) \times U(1)$ become the longitudinal components of the W and Z gauge bosons. But only the Goldstone bosons are directly involved — not the Higgs boson. The Goldstone bosons are always there whenever a continuous symmetry is spontaneously broken. But the existence of the Higgs boson depends on the dynamics of symmetry breaking. While the simplest weak-coupling versions of this dynamics (that we described in chapter 2) involves a fundamental Higgs boson, we will argue later that in some strong-coupling models of symmetry breaking, there is no Higgs boson at all, while in others, the Higgs boson exists, but is a composite state.

8.2 Effective Field Theories

Before we try to apply the results of the previous section to actual calculations, we will introduce some more machinery, the idea of an effective field theory. We have already used parts of the idea in Chapters 5 and 6 on chiral Lagrangians. There we used it in desperation and ignorance in order to concentrate on that part of the physics that we could hope to understand. Beginning with this section, we will sharpen the idea to the point where it will become an extremely useful calculational tool. It will make hard calculations easy and impossible calculations doable. In fact, however, the effective–field–theory idea is much more. It is one of the great unifying concepts in modern quantum field theory. Learn the language and it will serve you well.

The effective–field–theory idea is important because physics involves particles with very disparate masses and because we study physics in experiments involving various energies. If we had to know everything about all the particles, no matter how heavy, we would never get anywhere. But we don't.

Quantum electrodynamics, for example, describes the properties of electrons and photons at energies of the order of 1 MeV or less pretty well, even if we ignore the muon, quarks, the weak interactions, and anything else that may be going on at high energy. This works because we can write an effective field theory involving only the electron field and the photon field. With a completely general quantum field theory involving these fields, including arbitrary renormalizable interactions, we could describe the most general possible interactions consistent with relativistic invariance, unitarity of the S–matrix, and other general properties like TCP symmetry. Thus, we do not give up any *descriptive* power by going to an effective theory.

It might seem that we have given up *predictive* power because an arbitrary effective theory has an infinite number of nonrenormalizable interactions and thus an infinite number of parameters. But this is not quite right for two reasons, one quantitative and one qualitative. Quantitatively, if we know the underlying theory at high energy, then we can calculate all the renormalizable interactions. Indeed, as we will discuss, there is a straightforward and useful technology for performing these calculations. Thus, quantitative calculations can be done in effective theory language.

The qualitative message is even more interesting. As we will see in detail, so long as the underlying theory makes sense, all of the nonrenormalizable interactions in the effective theory are due to the heavy particles. Because of this, the dimensional parameters that appear in the nonrenormalizable interactions in the effective theory are determined by the heavy particle masses. If these masses are all very large compared to the electron mass and the photon and electron energies, the effects of the nonrenormalizable interactions will be small, suppressed by powers of the small mass or momenta over the large masses.

Thus, not only do we not lose any quantitative information by going to the effective–field–theory language, but we gain an important qualitative insight. When the heavy particle masses are large, the effective theory is approximately renormalizable. It is this feature that explains the success of renormalizable QED.

To extract the maximum amount of information from the effective theory with the minimum effort, we will renormalize the theory to minimize the logarithms that appear in perturbation theory. In practice, we will use a mass–independent renormalization scheme such as the $\overline{MS}$ scheme, and choose the renormalization scale μ appropriately. If all the momenta in a process of interest are of order μ, there will be no large logarithms in perturbation theory. The standard technology of the renormalization group can be used to change from one μ to another.

In the extreme version of the effective–field–theory language, we can associate each particle mass with a boundary between two effective theories. For momenta less than the particle mass, the corresponding field is omitted from the effective theory. For larger momenta, the field is included. The connection between the parameters in the effective theories on either side of the boundary is now rather obvious. We must relate them so that the description of the physics just below the boundary (where no heavy particles can be produced) is the same in the two effective theories. In lowest order, this condition is simply that the coupling constants for the interactions involving the light fields are continuous across the boundary. Heavy–particle exchange and loop effects introduce corrections as well as new nonrenormalizable interactions. The relations between the couplings imposed by the requirement that the two effective theories describe the same physics are called "matching conditions". The matching conditions are evaluated with the renormalization scale μ in both theories of the order of the boundary mass to eliminate large logarithms.

If we had a complete renormalizable theory at high energy, we could work our way down to the effective theory appropriate at any lower energy in a totally systematic way. Starting with the mass M of the heaviest particles in the theory, we could set $\mu = M$ and calculate the matching conditions for the parameters describing the effective theory with the heaviest particles omitted. Then we could use the renormalization group to scale μ down to the mass M' of the next heaviest particles. Then we would match onto the next effective theory with these particles omitted and then use the renormalization group again to scale μ down further, and so on. In this way, we obtain a descending sequence of effective theories, each one with fewer fields and more small renormalizable interactions than the last. We will discuss several examples of this procedure in the sections to come.

There is another way of looking at it, however, that corresponds more closely to what we actually do in physics. We can look at this sequence of effective theories from the bottom up. In this view, we do not know what the renormalizable theory at high energy is, or even that it exists at all. We can replace the requirement of renormalizability with a condition on the nonrenormalizable terms in the effective theories. In the effective theory that describes physics at a scale μ, all the nonrenormalizable interactions must have dimensional couplings less than $1/\mu$ to the appropriate power ($1/\mu^{D-4}$ for operators of dimension D). If there are nonrenormalizable interactions with coupling $1/M$ to a power, for some mass $M > \mu$, there must exist heavy particles with mass $m \lesssim M$ that produce them, so that in the effective theory including these particles, the nonrenormalizable interactions disappear. Thus, as we go up in energy scale in the tower of effective field theories, the effects of nonrenormalizable interactions grow and become important on the boundaries between theories, at which point they are replace by renormalizable (or, at least, nonrenormalizable) interactions involving heavy particles.

The condition on the effective theories is probably a weaker condition than renormalizability. We can imagine, I suppose, that this tower of effective theories goes up to arbitrarily high energies in a kind of infinite regression. This is a peculiar scenario in which there is really no complete theory of physics, just a series of layers without end. More likely, the series does terminate, either because we eventually come to the final renormalizable theory of the world, or (most plausible) because at some very large energy scale (the Planck mass?) the laws of relativistic quantum field theory break down and an effective quantum field theory is no longer adequate to describe physics.

Whatever happens at high energies, it doesn't effect what we actually do to study the low–energy theory. This is the great beauty of the effective–field–theory language.

8.3 The Symmetries of Strong and Electroweak Interactions

As a trivial example of the utility of the effective Lagrangian language, we will interpret and answer the following question: What are the symmetries of the strong and electromagnetic interactions? This question requires some interpretation because the strong and electromagnetic do not exist in isolation. Certainly the electromagnetic interactions, and probably the strong interactions as well, are integral parts of larger theories that violate flavor symmetries, parity, charge conjugation, and so on. The effective–field–theory language allows us to define the question in a very natural way. If we look at the effective theory at a momentum scale below the W and Higgs mass, only the leptons, quarks, gluons, and the photon appear. The only possible renormalizable interactions of these fields consistent with the $SU(3) \times U(1)$ gauge symmetry are the gauge invariant QED and QCD interactions with arbitrary mass terms for the leptons and quarks. All other interactions must be due to nonrenormalizable couplings in the effective theory and are suppress at least by powers of $1/M_W$. Thus it makes sense to define what we mean by the strong and electromagnetic interactions as the QED and QCD interactions in the effective theory. Indeed, this definition is very reasonable. It is equivalent to saying that the strong and electromagnetic interactions are what is left when the weak interactions (and any weaker interactions) are turned off (by taking $M_W \to \infty$).

Now that we have asked the question properly, the answer is rather straightforward. If there are n doublets of quarks and leptons, the classical symmetries of the gauge interactions are an $SU(n) \times U(1)$ for each chiral component of each type of quark. These $SU(n)$ symmetries of the gauge interactions do not depend on any assumptions about the flavor structure of the larger theory. They are automatic properties of the effective theory.

Now the most general mass terms consistent with $SU(3) \times U(1)$ gauge symmetry are arbitrary mass matrices for the charge $\frac{2}{3}$ quarks, the charge $-\frac{1}{3}$ quarks, and the charged leptons and an arbitrary Majorana mass matrix for the neutrinos.

The Majorana neutrino masses, if they are present, violate lepton number conservation. In fact, such masses have not been seen. If they exist, they must by very small. It is not possible to explain this entirely in the context of the effective $SU(3) \times U(1)$ theory. However, if we go up in scale and look at $SU(3) \times SU(2) \times U(1)$ symmetric theory above M_W and M_Z and if the only fields are the usual fermion fields, the gauge fields, and $SU(2)$ doublet Higgs fields, the renormalizable interactions in the theory automatically conserve lepton number.[1] Thus, there cannot be any neutrino mass terms induced by these interactions. If neutrino mass terms exist, they must be

[1] Classically, they conserve lepton number and baryon number separately. The quantum effects of the $SU(2) \times U(1)$ instantons produce very weak interactions that violate lepton number and baryon number but conserve the difference. Conservation of $B - L$ is enough to forbid neutrino masses.

suppressed by powers of an even larger mass. Presumable, this is the reason they are so small.

The quark and lepton masses can be diagonalized by appropriate $SU(n)$ transformations on the chiral fields. Thus, flavor quantum numbers are automatically conserved. Strong and electromagnetic interactions, for example, never change a d quark into an s quark. This statement is subtle, however. It does not mean that the flavor–changing $SU(2) \times U(1)$ interactions have no effect on the quark masses. It does mean that any effect that is not suppressed by powers of M_W can only amount to a renormalization of the quark fields or mass matrix, and thus any flavor–changing effect is illusory and can be removed by redefining the fields.

With arbitrary $SU(n)$ transformations on the chiral fields, we can make the quark mass matrix diagonal and require that all the entries have a common phase. We used to think that we could remove this common phase with a chiral $U(1)$ transformation so that the effective QCD theory would automatically conserve P and CP. We now know that because of quantum effects such as instantons this $U(1)$ is not a symmetry of QCD. A chiral transformation changes the θ parameter defined in (5.5.2), Thus, for example, we can make a chiral $U(1)$ transformation to make the quark masses completely real. In this basis, the θ parameter will have some value. Call this value $\overline{\theta}$. If $\overline{\theta} = 0$, QCD conserves parity and CP. But if $\overline{\theta} \neq 0$, it does not.

Experimentally, we know that CP is at least a very good approximate symmetry of QCD. The best evidence for this comes from the very strong bound the neutron electric dipole moment, less than $6 \times 10^{-25} e$ cm and going down. This implies that $\overline{\theta}$ is less than about 10^{-9}, a disturbingly small number. This situation is sometimes referred to as the strong CP puzzle. The puzzle is: why is $\overline{\theta}$ so small if the underlying theory violates CP?

8.4 The ρ Parameter

An interesting and important application of effective field theory ideas arises because the t quark is very heavy. It is observed in high-energy collider experiments to have a mass nearly twice that of the W and Z, about 175 GeV. Since the heavy t is not present in the effective theory below M_W, its only possible effect would be to change the structure of the theory at the W scale.

At this point, it is important to distinguish between fermions like the quarks and leptons that get mass due to their couplings to the Higgs field and fermions that have $SU(2) \times U(1)$ invariant masses. The latter might include, for example, a heavy Dirac doublet, in which both the left- and right-handed fields are $SU(2)$ doublets. Such heavy particles do not affect the low–energy theory very much. Their $SU(2) \times U(1)$ invariant masses ensure that they are removed from the effective theory in complete, degenerate $SU(2) \times U(1)$ multiplets. This does not affect the structure of the low–energy $SU(2) \times U(1)$ invariant theory, except to renormalize parameters and produce nonrenormalizable terms that are suppressed by powers of the large mass. Certainly, there are no effects of such particles that grow as their masses increase.

The situation is entirely different for the t quark or other particles with $SU(2) \times U(1)$–breaking mass terms. Below m_t, but above M_Z, the theory looks very peculiar. The $SU(2) \times U(1)$ gauge invariance is explicitly broken in the couplings of the W and Z to quarks because the t has been removed from the (t, b) doublet. And because the gauge invariance is broken, the other couplings in the effective theory are not constrained by it!

The most interesting terms in the effective Lagrangian from this point of view are the mass terms for the $W^{\pm}$ and W_3 gauge bosons that determine the relative strength of neutral– and charged–current weak interactions. They are interesting because they are dimension–2 operators and their coefficients can be proportional to the large m_t^2. A term that is the same for $W^{\pm}$ and W_3 is not interesting because it can be absorbed into a renormalization of the VEV of the Higgs field.

But a difference between the $W^{\pm}$ and W_3 mass term is a real physical effect. When the t quark is removed from the theory, exactly such an effect is produced in the matching condition onto the theory below m_t. The difference is

$$m_{W^\pm}^2 - m_{W_3}^2 = \delta M_W^2 \equiv \frac{3\alpha}{16\pi \sin^2\theta}\, m_t^2\,. \tag{8.4.1}$$

This is an effect of the heavy particle that actually grows with the heavy mass! It is easiest to interpret (8.4.1) as an additional correction to the $W^{\pm}$ mass, because the W_3 mass term gets mixed up with the X mass in the spontaneous symmetry breaking, so this is what we will do.

Another way to understand (8.4.1) is to notice that since m_t arises from a Yukawa coupling

$$f_t = \frac{\sqrt{2}\, m_t}{v}, \tag{8.4.2}$$

we can write (8.4.1) as

$$\frac{m_{W^\pm}^2 - m_{W_3}^2}{m_{W^\pm}^3} = \frac{3 f_t^2}{32\pi^2}\,. \tag{8.4.3}$$

This shows that the effect is a radiative correction that depends on the Yukawa coupling.

(8.4.1) is phenomenologically important (as well as theoretically interesting) not only because it affects the W and Z masses, but because it changes the relative strength of neutral– to charged–current process. It is conventional to define a parameter ρ that is the measured ratio of the coefficient of $j_3^\alpha j_{3\alpha}$ to $j_{1,2}^\alpha j_{1,2\alpha}$ (see (2.8.6)) in the effective weak Hamiltonian for neutrino interactions. ρ is known to be very close to 1 in neutrino–hadron scattering. Actually, we there is a negative contribution to ρ because of the QED renormalization of charged–current scattering (Figure 8–4 and (8.6.2)). The heavy t gives a positive contribution to ρ (in all processes):

$$\Delta\rho = \frac{3\alpha}{16\pi \sin^2\theta}\, \frac{m_t^2}{M_W^2}. \tag{8.4.4}$$

It is interesting to note that a doublet (t', b') in which the t' and b' were heavy and *degenerate*, there would be no such effect. Removing the t' and b' from the theory produces a common renormalization of $m_{W^\pm}$ and m_{W_3} that is not physically interesting. Even though the t and b masses break the $SU(2) \times U(1)$ symmetry, they do not break the custodial $SU(2)$ symmetry see (2.8.7) that keeps $\rho = 1$. However, there are other interesting effects of this doublet that grow like $\ln(m_{t',b'}/M_W)$. We will return to these below.

The analysis of the ρ parameter was done (in a different language) by M. B. Einhorn, D. R. T. Jones, and M. Veltman (*Nuclear Physics* **B191**, 146–172, 1981), who also noticed that contributions to ρ of this kind all have the same sign, no matter how the heavy particles transform. This puts an interesting constraint on possible heavy weakly interacting matter.

8.5 M_W and M_Z

In Chapter 2, we learned that the W and Z masses are predicted in the $SU(2) \times U(1)$ model. In the tree approximation, we find from (2.8.1) and (2.7.28)

$$M_W^2 = \frac{\sqrt{2}\, e^2}{8 G_F \sin^2\theta}, \qquad M_Z = M_W / \cos\theta. \tag{8.5.1}$$

We have discussed the experiments that determine the parameters in (8.5.1). We know from QED that

$$\alpha = \frac{e^2}{4\pi} = \frac{1}{137.036} \tag{8.5.2}$$

We can determine G_F from the μ decay rate,

$$\tau_\mu^{-1} = \frac{G_F^2 m_\mu^5}{192\pi^3}, \tag{8.5.3}$$

which for measured τ_μ and m_μ gives

$$G_F = 1.164 \times 10^{-5} \text{ GeV}^{-2}. \tag{8.5.4}$$

Finally, $\sin^2\theta$ as measured in neutral–current experiments (from a more sophisticated version of (7.5.5), is

$$\sin^2\theta \simeq 0.23 \tag{8.5.5}$$

Putting (8.5.2), (8.5.4), and (8.5.5) into (8.5.1) gives

$$M_W = 77.8 \text{ GeV}, \quad M_Z = 88.7 \text{ GeV}. \tag{8.5.6}$$

The tree level predictions, (8.5.6), are significantly off from the measured values in the particle data book:

$$M_W = 80.33 \text{ GeV}, \quad M_Z = 91.187 \text{ GeV}. \tag{8.5.7}$$

We would like to calculate the leading radiative corrections to (8.5.6) to see whether they account for this discrepancy. By far the easiest way to do this is to adopt an effective–field–theory language. One important correction is the contribution from the heavy t that we discussed in the previous sections. The other dominant effects can be understood as follows. Below M_W the effective theory involves QCD and QED, with the weak interactions appearing only as nonrenormalizable four–fermion interactions. In lowest order, the matching condition that determines these four–fermion interactions comes from single W and Z exchanges. That means that in leading order, G_F, e^2, and $\sin^2\theta$ should be interpreted as parameters renormalized at $\mu \simeq M_W$ and the coefficients of the various four–fermion operators are given by the tree–approximation formulae discussed in Chapters 2 and 3 (including the effect of the t quark on the W mass). Higher–order contributions will be a power of series in the $SU(2)$ and $U(1)$ coupling constants, of order

$$\frac{\alpha}{4\pi\,\sin^2\theta}, \quad \frac{\alpha}{4\pi\,\cos^2\theta}. \tag{8.5.8}$$

These effects are less than 1%.

There are, however, much more important effects. The most important is that the α that appears in (8.5.1) is not given by (7.4.2). It should be interpreted as $\alpha(M_W)$, where $\alpha(\mu)$ is the running coupling constant in the effective QED theory renormalized at μ. The other effects are similar. The effective four–fermion operators obtained from the matching condition are renormalized at M_W, but the experiments that are used to determine the parameters are done at smaller momentum. The renormalization group must be used to find the form of four–fermion operators renormalized at the μ appropriate for each experiment.

Actually, the renormalization group is an affectation in these calculations. The important point is that these corrections involve large logarithms,

$$\frac{\alpha}{4\pi} \ln \frac{M_W}{\mu}, \tag{8.5.9}$$

and that all flavors of quarks and charged leptons give contributions. The renormalization group automatically adds up the higher powers of $\alpha \ln M_W/\mu$, but here α is small enough to make the higher–order terms negligible. The terms proportional to (8.5.9) can be extracted directly from one–loop diagrams.

We will first the renormalization of α. The conventional α given in (8.5.2) is renormalized on the electron mass shell. However, the difference between this definition and a reasonable running coupling (like the MS scheme) evaluated with $\mu = m_e$ is not very large because it has no large logarithm in it. Thus, to a good approximation, we can take

$$\alpha(m_e) = 1/137. \tag{8.5.10}$$

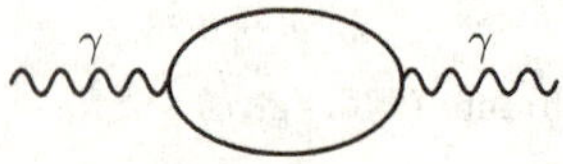

Figure 8-1:

Now we can follow $\alpha(\mu)$ up to $\mu \simeq M_W$. The μ dependence comes from the vacuum polarization diagram, as shown in Figure 8–1. All charged particles that have masses less than μ contribute in this process (Figure 8–1) because these are the charged particles in the effective QED theory at the scale μ. A standard calculation gives

$$\alpha(\mu_2) = \alpha(\mu_1) \cdot \left[1 + \frac{2\alpha}{3\pi} \sum_i Q_i^2 \ln \frac{\mu_2}{\mu_1}\right], \tag{8.5.11}$$

where the sum runs over all light fermions in the effective theory. Thus, we find

$$\begin{aligned} \alpha(M_W) &= \alpha(m_e) \left[1 + \frac{2\alpha}{3\pi} \sum_i Q_i^2 \ln \frac{M_W}{m_i}\right] \\ &\simeq 1/129. \end{aligned} \tag{8.5.12}$$

Here I have used constituent masses for the quarks (because quarks should not be in the effective theory for momenta much smaller than the mass of the bound states containing those quarks). This is a 6% increase in $\alpha(M_W)$ compared to (8.5.2), which translates to a 3% increase in M_W and M_Z in (8.5.1).

The Fermi constant G_F is determined from the μ decay rate, so we must look at the renormalization of the four–fermion operator

$$\frac{G_F}{\sqrt{2}} \overline{\nu}_e \gamma^\mu (1+\gamma_5) e^- \overline{u^-} \gamma_\mu (1+\gamma_5) \nu_\mu. \tag{8.5.13}$$

For convenience, we will work in Landau gauge where the fermion–wave–function renormalization from the diagram seen in Figure 8–2 has no μ dependence. Then the renormalization of (8.5.13) comes only from the diagram in Figure 8–3 where the arrows indicate the direction in which the

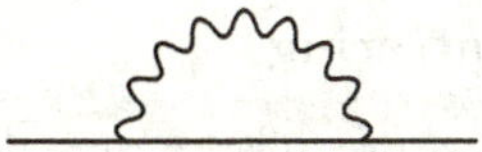

Figure 8-2:

lepton number "flows". That is, the e^- line with the arrow pointing towards the vertex means either an incoming electron or outgoing positron. In this case, because only left-handed fields are involved in (8.5.13), the arrows could also indicate the flow of left-handedness. The photon can couple only as shown because the neutrinos have no electric charge. But the diagram in Figure 8–3 gives no $\ln\mu$ dependence. The diagram is finite.

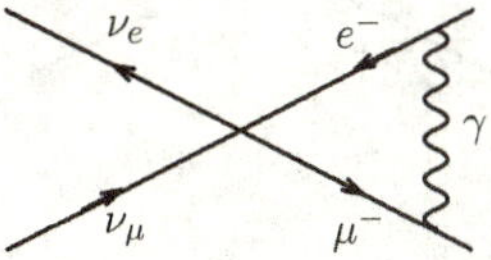

Figure 8-3:

We can see this by direct calculation or note that we could have used Fierz transformations (Problem 2–3) to write (8.5.13) in so-called charge retention form:

$$\frac{G_F}{\sqrt{2}}\overline{\nu}_e\gamma^\mu(1+\gamma_5)\nu_\mu\overline{\mu^-}\gamma_m u(1+\gamma_5)e^- . \tag{8.5.14}$$

Here it is clear that the μ and e fields appear only in the form of the left chiral current, which is conserved in the limit that the μ and e masses vanish. But a conserved current does not get multiplicatively renormalized. It is related by Noether's theorem to a charge that is a physical observable.

Thus, there are no corrections of order $\alpha\ln M_W/\mu$ to (8.5.13). That means that the G_F determined by the μ decay rate is the same as the G_F that appears in (8.5.1).

We can now put these corrections together to get one test of the standard model that incorporates the leading radiative corrections. We will use the measured values of M_Z, m_t and G_F, and the corrected $\alpha(M_W)$ of (8.5.12) to predict M_W. We begin by incorporating the corrections we have discussed, into (8.5.1), and using this to determine $\sin^2\theta$:

$$\cos^2\theta M_Z^2 + \delta M_W^2 = \frac{\sqrt{2}\,e^2}{8G_F\sin^2\theta}, \cos\theta. \tag{8.5.15}$$

Solving for $\sin^2\theta$ gives

$$\sin^2\theta = 0.228 \tag{8.5.16}$$

Then

$$M_W = M_Z\cos\theta = 80.1\ \text{GeV} \tag{8.5.17}$$

which agrees with the measured value to within a fraction of a percent!

8.6 Neutrino-hadron scattering

In this section, we will briefly discuss the radiative corrections to low-energy neutrino hadron scattering. The operator that contributes to charged–current neutrino hadron scattering is

$$\frac{G_F}{\sqrt{2}}\overline{\nu^-}\gamma^\mu(1+\gamma_5)\nu_\mu\overline{U}\gamma_\mu(1+\gamma_5)D. \tag{8.6.1}$$

It gets renormalized because of the diagram in Figure 8–4 that gives for the coefficient

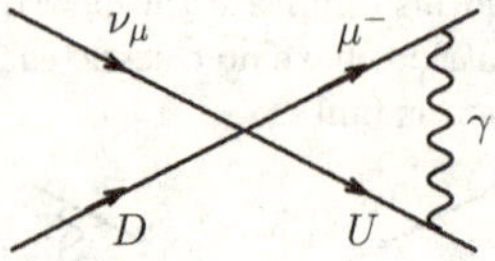

Figure 8-4:

$$C(\mu) = \frac{G_F}{\sqrt{2}}\left(1+\frac{\alpha}{\pi}\ln\frac{M_W}{\mu}\right). \tag{8.6.2}$$

Note that there is an effect only when the photon is exchanged between legs with the same outgoing handedness. Otherwise, as discussed above, there is no large logarithmic renormalization.

The four–fermion operator relevant in neutral–current neutrino scattering is

$$\frac{G_F}{\sqrt{2}}\overline{\nu}_f\gamma^\mu(1+\gamma_5)\nu_f$$
$$\left(\sum_{j\neq f}\overline{\psi}_j\gamma_\mu\left(T_3(1+\gamma_5)-2\sin^2\theta Q\right)\psi_j+\overline{f}\gamma_\mu\left(\frac{1}{2}(1+\gamma_5)+2\sin^2\theta\right)f\right) \tag{8.6.3}$$

where the sum runs over all flavors of leptons and quarks, and the f refers to the lepton flavor, e, μ or τ. The f is necessary because there is a contribution from W exchange as well as the contribution from Z exchange (see problem (2-5)). This is like (8.5.13) in that there is no multiplicative renormalization. Here, however, there is a new effect produced by the diagram in Figure 8–5.

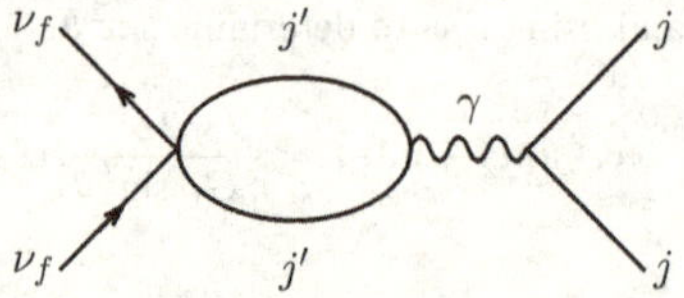

Figure 8-5:

One might think (for a fleeting moment) that the photon operator would produce at pole for momentum transfer $q^2 = 0$, which would give rise to a long–range neutral–current momentum interaction. This cannot happen, because the subdiagram shown in Figure 8–6, where x is the electromagnetic current, must vanish as the momentum transfer goes to zero because the neutrino has zero electric charge. Thus, in (8.6.3) the Feynman integral produces a factor of q^2 that kills the pole.

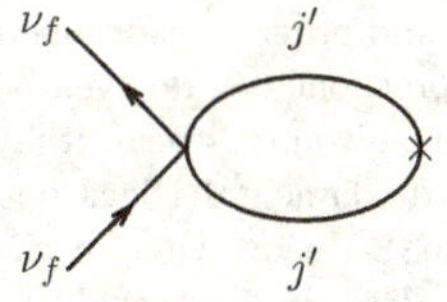

Figure 8-6:

The result is a four-fermion interaction that is a product of the neutrino current times the electromagnetic current. In other words, it simply renormalizes $\sin^2\theta$ in (8.6.3):

$$\sin^2\theta = \sin^2\theta M_W$$

$$\cdot\left[1+\frac{\alpha}{3\pi}\left(\sum_{j\neq f} Q_j\left(T_3\frac{1}{\sin^2\theta_{M_W}}-2Q_j\right)\ln M_W/\max(\mu,\, m_j)\right.\right. \tag{8.6.4}$$

$$\left.\left.-\left(\frac{1}{2\sin^2\theta_{M_W}}+2\right)\ln M_W/\max(\mu,\, m_f)\right)\right]$$

The renormalization group interpretation of (8.6.4) is slightly subtle. The diagram (Figure 8–5) is not one–particle irreducible, so it would not normally be included in the analysis of operator mixing. In fact, the relevant diagram is seen in Figure 8–7, which produces mixing between (8.6.3) and the dimension–6 operator

$$\frac{1}{e}\overline{\nu}\gamma^\mu(1+\gamma_5)\nu\partial^\lambda F_{\lambda\mu}. \tag{8.6.5}$$

But $\partial^\lambda F_{\lambda\mu}/e$ is related to the electromagnetic current by the equations of motion. In general, it is not true that the equations of motion can be used to replace one operator with another. The classical equations of motion are not as true as operator statements. But we can use the equations of motion to evaluate physical, on–shell matrix elements of the operators.

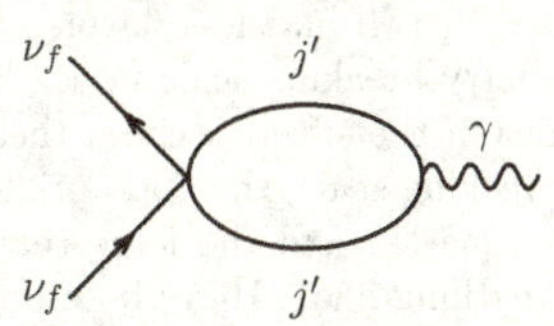

Figure 8-7:

The net effect of the corrections to the four-fermion operators for neutrino–hadron scattering is to reduce the value of $\sin^2\theta$ relevant in (8.5.1). The neutral–current expression (8.6.4) reduces $\sin^2\theta$ by a few percent if the experiments that determine $\sin^2\theta$ have q^2 of a few Gev^2. The result agrees well with the value of $\sin^2\theta$ extracted from M_Z in the previous section.

8.7 Technicolor

We have seen that the leading radiative corrections to the standard model give rather good agreement with the measured values of the W and Z masses and the weak mixing angle. We could go on and calculate W and Z partial width and other measurable quantities with similar success. As discussed in the introduction to this chapter, one can do even better by doing a complete analysis of all the radiative corrections, and the result agrees beautifully with the simplest version of the standard model, with a relatively light, fundamental Higgs boson (for a review available on the web, see G. Altarelli *etal*, hep-ph/9712368). In my view, however, one should be wary of interpreting this success as evidence that the Higgs boson actually exists. The really important moral of the radiative corrections to the standard model is precisely that the most important radiative corrections are independent of most of the details of electroweak symmetry breaking. In fact, in our analysis above, the only really important constraint on this physics is that it must preserve a custodial $SU(2)$ symmetry in the limit that the $U(1)$ gauge coupling is turned off. Otherwise, we would not be able rule out large corrections to the relation $M_W = M_Z \cos\theta$. All other contributions from the physics of symmetry breaking are generically of the size of matching corrections to the low energy theory below the weak scale, at the level of fractions of a percent.

Because the physics of electroweak symmetry breaking is so difficult to see in the low energy physics of the standard model, it is very useful to separate the two and define what we mean by the standard model to be the physics of the quarks and leptons, the $SU(3) \times SU(2) \times U(1)$ gauge interactions, and the Goldstone bosons that are eaten to become the longitudinal components of the W and Z. As we have seen, essentially the only thing we have to know about the Goldstone bosons is that the spontaneous breaking that produces them preserves a custodial $SU(2)$ symmetry. Then the standard model emerges, including the large radiative corrections. In this language, we distinguish between the "Higgs mechanism" and the "Higgs boson." The Higgs mechanism is a crucial part of the standard model. It is the process by which the Goldstone bosons that must arise because of the Goldstone theorem from the spontaneous breaking of $SU(2) \times U(1)$ combine with the gauge degrees of freedom to become the massive W and Z. We see these Goldstone bosons when we produce the W and Z. The Higgs boson, on the other hand, is still a hypothetical particle. Whether or not anything like a Higgs boson exists depends on what the Goldstone bosons are actually made of. If the Goldstone bosons are fundamental particles, then they must be part of Higgs multiplet that transforms linearly under the $SU(2) \times U(1)$ symmetry. In this case, the Higgs boson exists, and is the partner of the Goldstone bosons. However, if the Goldstone bosons are composite states, like the pions in QCD, then there may be nothing like a Higgs boson at all.

Consider the situation in QCD with two massless flavors. The Goldstone bosons appear in the effective theory below the symmetry breaking scale in a 2×2 version of the U field that we used to describe the pion interactions in the effective chiral theory in chapter 5. They transform nonlinearly. Above the symmetry breaking scale, the appropriate language to describe the theory is the language of the constituents (quarks) and the force that binds them (QCD and gluons). Neither the Goldstone bosons nor anything like a Higgs boson appears in the high energy theory. Note that in QCD, there is an analog of custodial $SU(2)$ — it is simply isospin symmetry.

One can imagine that a similar thing happens for the electroweak interaction. If the Goldstone bosons of $SU(2) \times U(1)$ breaking are bound states of constituent "techniquarks" bound by a very strong gauge interaction, "technicolor", then we can describe the low energy theory in terms of a 2×2 unitary U field that contains the Goldstone bosons. It is very easy to see that in leading order in the momentum expansion, this exactly reproduces the standard model masses and couplings of the gauge bosons, so long as the technicolor dynamics preserves a custodial $SU(2)$. This follows simply from the discussion following (2.8.7), where we showed that the Higgs doublet in Weinberg's original model of leptons could be rewritten in terms of a 2×2 Σ field, and that the theory then automatically has a custodial $SU(2)$ symmetry when the $U(1)$ gauge interactions are turned off.

The relation between the Higgs doublet, Σ, and the U field is simply

$$\Sigma = (v + h)\, U \tag{8.7.1}$$

where v is the vacuum expectation value of Σ, and h is shifted field associated with the Higgs boson. Thus if we ignore h, the standard model kinetic energy of the Higgs doublet reduces to the leading kinetic energy term for the Goldstone bosons.

The difficulty with technicolor models is not with the W and Z properties, which are well described in a technicolor theory, but with the generation of masses for the matter particles. In a model with a fundamental Higgs doublet, and thus a Higgs boson, these masses arise from the Yukawa couplings of the Higgs doublet to the quarks and leptons, which is a renormalizable interaction. No one has found a totally convincing alternative mechanism for mass generation in a model without fundamental scalar fields. It is not clear whether this is a problem with technicolor models, or simply the lack of imagination of theorists. It seems to be impossible to build models based on simple rescaled versions of QCD, for a variety of reasons. But it is entirely possible that Nature has other tricks up her sleeve.

Finally, it is worth mentioning that there is an intermediate possibility, in which the Higgs boson exists, but is a composite state (see D. Kaplan and H. Georgi, Phys. Lett. **136B**, 183 (1984)). A particularly interesting possibility is the idea that the constituents of the Higgs might include the t quark. Work in this direction is continuing (see for example, B.A. Dobrescu and C.T. Hill, hep-ph/9712319).

Problems

8-1. Derive (8.1.14).

8-2. Derive (8.1.20) and (8.1.21) from (8.1.19).

8-3. Write a nonrenormalizable $SU(2)\times U(1)$ invariant interaction between lepton doublets and Higgs doublets that gives a Majorana mass to the neutrino after spontaneous symmetry breaking.

8-4. Calculate the diagrams in Figures 8-1, 8-4 and 8-5.

9 — Nonleptonic Weak Interactions

9.1 Why We Can't Calculate

Thus far, I have not said very much about the nonleptonic weak decays of light hadrons. The reason is that theoretical understanding of the nonleptonic weak interactions is rather limited. We saw in Chapter 7 that we can calculate certain interesting quantities in the decay of a state containing a sufficiently heavy quark using the parton model, but we cannot discuss individual modes in this way. For strange–particle nonleptonic decays, we can say very little that is quantitative, apart from a few relations between different modes obtained from chiral Lagrangians or current algebra considerations (see Section 5.10).

The reason for our ignorance is simple. We just do not understand the low–energy strong QCD interactions in quantitative detail. Most of our understanding is based on symmetry arguments, as in the chiral Lagrangian techniques of Chapters 5 and 6. Even when supplemented by dynamical guesses, as in the chiral–quark model of 6.4, these techniques cannot be used to reliably calculate nonleptonic decay rates. I will make this argument very explicit at the end of this section.

To begin with, I will formulate the discussion of nonleptonic decays in the effective Lagrangian language. This will probably be the most convenient way to do calculations if anyone ever does actually achieve a quantitative understanding of strong interactions. Like good scouts, we will be prepared. The more serious reason for the effective Lagrangian analysis is that it allows us to divide the problem into two parts, the first of which we understand. This will enable us to make interesting qualitative statements, even without a detailed quantitative understanding.

The exchange of virtual W produces a variety of four–quark operators in the effective Lagrangian just below M_W. All we have to do to construct the effective Lagrangian that is relevant at low energies is to use the renormalization group to bring down the renormalization scale μ from M_W to the scale of interest. This produces an effective low–energy Lagrangian in which the original four–quark operators appear along with other dimension–6 operators that are produced by the renormalization group. Each coefficient (a coupling constant in the effective theory) is given explicitly by the renormalization group in terms of the QCD coupling constant and the W and quark masses. This is the part of the problem that we understand.

To use the effective Lagrangian to actually calculate a physical decay rate, we must calculate the matrix element of the effective Hamiltonian between the relevant physical states. This is the hard part. We do not know how to calculate the matrix elements of operators between physical hadron states, even when the operators are renormalized at the appropriate small momentum scale.

We might try to calculate the matrix elements of the appropriate operators by appealing to the chiral Lagrangian arguments of Chapter 5. This was very useful in the study of leptonic and semileptonic decays. But here it doesn't help. In the effective Lagrangian for semileptonic charged–current interactions, the part involving quarks is a flavor current. Thus, we know the corresponding operator in the effective chiral Lagrangian by pure symmetry arguments. Both the form and the normalization of the current operator are determined by the symmetry. But the nonleptonic operators are not currents. They look like products of currents at M_W, but all that tells us is that they transform in a particular way under the flavor symmetries. We must choose the corresponding operator in the chiral theory to transform the same way. But that only constrains the form of the chiral operator. It tells us nothing about the normalization.

One might think that the situation would be better in a chiral–quark model. The quark fields appear explicitly, so perhaps we can simply take over four–quark operators directly into the effective chiral–quark Lagrangian. But this is cheating! In going from the QCD theory with light quarks to the chiral theory with constituent mass quarks, we have changed the description of the theory, and we must calculate the matching conditions. These can change the form and normalization of the weak Hamiltonian. Indeed, we have seen an example of this in Section 6.5 in the renormalization of the axial vector current for chiral quarks.

Although we don't understand in detail how to do the calculations, it seems reasonable to suppose that the renormalizations induced by the matching conditions in going to the chiral–quark model are given by a power series in $\alpha_s(\Lambda_{\text{CSB}})$. This parameter could be small enough to make the renormalization factors manageable, like the 0.75 for g_A, but there will clearly be large uncertainties remaining.

Another promising direction in the calculation of these matrix elements is the development of numerical methods in lattice gauge theories. Eventually the techniques developed there may lead to direct evaluation of the weak interaction matrix elements. But don't hold your breath.

For now we will content ourselves with a calculation of the coefficients, the coupling constants of the effective theory. These already contain some interesting physics.

9.2 The Renormalization Group

Here I give a quick review of the renormalization group. We start by considering an effective theory with many couplings g_i with various physical dimensions. Now consider a set of physical quantities g_i^0. These may be scattering cross sections or particle masses. Or they may be more derived quantities, such as the (suitably regulated) bare parameters in the Lagrangian. At any rate, in perturbation theory, the quantities are functions G_i^0 (defined as a power series) of an arbitrary renormalization scale μ and the couplings renormalized at μ, which we denote by $g_i(\mu)$,

$$G_i^0\,(g(\mu),\,\mu)\,. \qquad (9.2.1)$$

Any physical quantity behaves this way, but I have chosen a set of physical quantities g_i^0 in one-to-one correspondence with the couplings g_i because I want to use the measured values of the g_i^0 to determine $g_i(\mu)$. This is easy. Just find $g_i(\mu)$ such that

$$g_i^0 = G_i^0\,(g(\mu),\,\mu)\,. \qquad (9.2.2)$$

These equations implicitly define the $g_i(\mu)$ that we can then use to calculate any physical quantity.

In principle, this works for any μ. But in practice, for a given set of physical quantities, some μ's will be more convenient than others. We want to choose μ to minimize the logarithms that appear in perturbation theory. That, after all, is the whole point. Perturbation theory will work best for the appropriate value of μ. After we have obtained $g_i(\mu)$, we can choose μ to make the calculation of any given physical quantity as convenient as possible. This also explains why we don't actually calculate $g_i(\mu)$ by solving (9.2.2), which would require calculating $G_i^0\,(g(\mu),\,\mu)$ accurately for all μ. Instead, we find differential equations for $g_i(\mu)$, which we can determine by calculating G_i^0 in the neighborhood of the most convenient μ.

We can write the μ dependence of $g_i(\mu)$ in differential form as

$$\mu \frac{d}{d\mu} g_i(\mu) = \beta_i\left(g(\mu),\, \mu\right). \tag{9.2.3}$$

To determine the β functions, we can differentiate (9.2.2) with respect to μ and use (9.2.3) to obtain

$$\left(\mu\frac{\partial}{\partial\mu} + \beta_i \frac{\partial}{\partial g_i}\right) G_i^0\,(g,\,\mu) = 0. \tag{9.2.4}$$

This is a renormalizable group equation. It expresses the fact that a change in the renormalization scale can be compensated by a corresponding change in the coupling constants, leaving physical quantities unchanged. (9.2.4) allows us to determine the functions β_i from the functional form of G_i^0. The important point is that to determine β_i, we need to know G_i^0 only in a neighborhood of a given μ.

It is often convenient to calculate the β_i by studying not physical quantities, but Green's functions or one–particle–irreducible functions. These also satisfy renormalization group equations. But here there are additional terms because of wave–function renormalization. The n $1PI$ function Γ^n for a multiplicatively renormalized field ϕ satisfies

$$\left(\mu\frac{\partial}{\partial\mu} + \beta_i \frac{\partial}{\partial g_i} - n\gamma_\phi\right)\Gamma^n = 0, \tag{9.2.5}$$

where γ_ϕ is the "anomalous dimension" of the ϕ field.

Composite operators, such as those that appear in the weak Hamiltonian, also have anomalous dimensions. However, in general, such operators are not multiplicatively renormalized but mix under renormalization. Suppose O_x for $x = 1$ to k is a set of operators that mix with one another under renormalization. Then we can write a renormalization group equation for an n point $1PI$ function with an o_x insertion Γ^n_xm as follows:

$$\left(\mu\frac{\partial}{\partial\mu} + \beta_i \frac{\partial}{\partial g_i} - n\gamma_\phi\right)\Gamma^n_x + \gamma_{xy}\Gamma^n_y = 0, \tag{9.2.6}$$

where γ_{xy} is the anomalous dimension matrix of the o_x operators. Note that the difference in the sign between the $n\gamma_\phi$ term and the γ_{xy} term is due to the fact that we are studying a $1PI$ function in which the ϕ legs have been amputated. In the renormalization group equation for the corresponding Green's functions, both would appear with a plus sign.

In general, β_i will depend explicitly on μ. But for a class of particularly convenient renormalization prescriptions, there is no explicit μ dependence. These are the "mass–independent" renormalization schemes, such as dimensional regularization with minimal subtraction. The important thing about these schemes is that the μ dependence is only logarithmic. Thus, $\beta_i(g)$ is a polynomial in the g's in which each term has the same physical dimension, the dimension of g_i.

In the full field theory of the world, a mass–independent renormalization scheme is rather useless. The problem is that heavy particles continue to influence the renormalization of coupling constants even at scales much smaller than their masses. This leads at such scales to large logarithms in the perturbation theory.

One alternative is to adopt a momentum subtraction scheme in which the scale μ is defined directly in terms of momenta. This avoids the theoretical difficulty of large logarithms at small scales, but it introduces a serious practical difficulty. The β functions have explicit μ dependence, and it is much harder to solve and use the renormalization group equations. Beyond one–loop order, it is practically impossible.

Fortunately, in the effective–field–theory language, a mass–independent renormalization scheme does not cause problems. The physical requirement that heavy particles must not contribute to renormalization at small scales is automatic because the heavy particles are simply removed from the effective theory. The combination of the effective–field–theory idea with mass–independent renormalization gives a scheme that is both physically sensible and practical. In what follows, we will always assume that we are working in such a scheme, although to lowest order it will not make very much difference. The real advantages of such schemes show up in two loops and beyond.

In the application of these ideas to the QCD corrections to the weak Hamiltonian, by far the most important coupling is the QCD coupling g. We will seldom be interested in going beyond first order in any other coupling constants. Thus, in the β function for g itself, we can ignore all other couplings and write

$$\mu\frac{\partial}{\partial\mu}g(\mu) = \beta(g(\mu)). \tag{9.2.7}$$

In lowest–order perturbation theory in QCD

$$\begin{aligned} \beta(g) &= -Bg^3 \\ B &= \tfrac{(11-2n/3)}{16\pi^2}. \end{aligned} \tag{9.2.8}$$

where n is the number of quarks in the effective theory. If we ignore higher–order terms in β and solve (9.2.7), we get

$$g(\mu)^{-2} = 2\beta \ln \mu/\Lambda, \tag{9.2.9}$$

or

$$\alpha_s(\mu) = \frac{2\pi}{(11-2n/3)\ln\mu/\Lambda}. \tag{9.2.10}$$

This holds in any one effective theory. To obtain the coupling in the effective theory at a given scale, we evaluate (9.2.10) with n equal to the number of quark species with mass smaller than μ and choose Λ to that α_s is continuous on the boundaries between regions. Thus, Λ changes from region to region. In terms of the Λ appropriate at low energies (for the three light quarks), we can write

$$\alpha_s(\mu) = 6\pi/\left(27\ln\mu/\Lambda - 2\sum_{m_q<\mu}\ln\mu/m_q\right), \tag{9.2.11}$$

which automatically incorporates matching conditions at the quark threshold.

Having determined $\alpha_s(\mu)$, we can now proceed to look at the weak Hamiltonian. The dominant contribution comes from dimension–6 operators, like the four–fermion operators induced by W exchange at M_W. At scales below M_W, the effective weak Hamiltonian is a linear combination of all of the operators that mix with the four–quark operators. Call these O_x. Then the effective weak Hamiltonian has the form

$$h_x(\mu)O_x, \tag{9.2.12}$$

where $h_x(\mu)$ are the couplings.

The β_x functions that describe the μ evolution of the $h_x(\mu)s$ must also have dimension m^{-2}. If we ignore quark masses (which is a sensible thing to do, because their effects are suppressed by powers of m_q/M_W), the only way that this can happen is to have β_x of the form

$$\beta_x(h,\, g) = \gamma_{xy}(g)h_y, \tag{9.2.13}$$

where the $\gamma_{xy}(g)$ are dimensionless functions of g. Such β functions describe the renormalization and mixing of the h_xs under the influence of QCD. To lowest order,

$$\gamma_{xy}(g) = A_{xy}g^2 + 0(g^4). \tag{9.2.14}$$

If we ignore the g^4 terms, it is easy to solve the evolution equations

$$\mu\frac{d}{d\mu}h_x(\mu) = A_{xy}g(\mu)^2 h_y(\mu). \tag{9.2.15}$$

If $A_\sim$ is the matrix with components A_{ij} and $h(\mu)$ is a column vector, we can write the solution as

$$h(\mu) = \exp\left\{\int_{\mu_0}^{\mu} Ag(\mu')^2\,\frac{d\mu'}{\mu'}\right\}h(\mu_0). \tag{9.2.16}$$

Because

$$g(\mu')^2 = -\frac{\beta(g(\mu'))}{Bg(\mu')} = -\frac{1}{Bg(\mu')}\,\mu'\frac{d}{d\mu'}g(\mu'), \tag{9.2.17}$$

we can do the integral in (9.2.11) and write

$$h(\mu) = \exp\left\{1(\underset{\sim}{A}/B\ln\left(\frac{g(\mu)}{g(\mu_0)}\right)\right\}h(\mu_0). \tag{9.2.18}$$

Linear combinations of the h's corresponding to eigenvalues of A are multiplicatively renormalized. That is, if ν is a fixed row vector such that

$$\nu\,\underset{\sim}{A} = a\nu, \tag{9.2.19}$$

then

$$\nu h(\mu) = (g(\mu)/g(\mu_0))^{-a/B}\nu h(\mu_0). \tag{9.2.20}$$

Of course, (9.2.17)–(9.2.20) are valid in any one effective theory. Appropriate matching conditions must be imposed at the boundaries.

9.3 Charm Decays

In the remainder of this chapter, we will ignore the mixing of the four light quarks with the b and t quarks. In this limit, as in a world with only four quarks, the weak interactions depend on only a single real parameter, the Cabibbo angle θ_c. The weak interactions do not violate CP. We will return and discuss CP violation in the general six–quark world in Chapter 10.

The simplest process from the point of view of the effective field theory is the $\Delta S = +1$ decay of the $\bar{c}$ antiquarks, for which the effective coupling to lowest order at the W scale is

$$\cos^2\theta_c\frac{G_F}{\sqrt{2}}\,\bar{c}\gamma^\mu(1+\gamma_5)s\bar{d}\gamma_\mu(1+\gamma_5)u. \tag{9.3.1}$$

The Hermitian conjugate is associated with c decay.

The renormalization of this coupling comes from the diagrams in which a gluon is exchanged between the $\bar{c}$ and $\bar{d}$ legs or the s and u legs. This is analogous to the electromagnetic renormalization of the charged–current semileptonic coupling that we discussed in Chapter 8 (see Figure 8–4), with photon exchange replaced by gluon exchange. Here, gluon exchange between legs with opposite handedness does not produce a logarithmic renormalization. Gluon exchange between legs of the same handedness increases the strength of the coupling as the scale decreases (that is, it gives a negative β function) if it produces an attractive force, and it decreases the strength of the coupling with decreasing scale for a repulsive force.

Gluon exchange in (9.3.1) is neither purely attractive nor purely repulsive because the fields of the same handedness are not in states of definite color. To simplify the analysis, we can form the combinations

$$\begin{aligned} 2O_{\pm}^{\overline{x}y} = &\ \overline{x}\gamma^{\mu}(1+\gamma_5)s\overline{d}\gamma_{\mu}(1+\gamma_5)y \\ &\pm\overline{x}\gamma^{\mu}(1+\gamma_5)y\overline{d}\gamma_{\mu}(1+\gamma_5)s, \end{aligned} \tag{9.3.2}$$

where x and y run over the various quark fields. Here we are interested in

$$\begin{aligned} 2O_{\pm}^{\overline{c}u} = &\ \overline{c}\gamma^{\mu}(1+\gamma_5)s\overline{d}\gamma_{\mu}(1+\gamma_5)u \\ &\pm\overline{c}\gamma^{\mu}(1+\gamma_5)u\overline{d}\gamma_{\mu}(1+\gamma_5)s. \end{aligned} \tag{9.3.3}$$

If we put in explicit color indices, α, $\beta = 1$ to 3, and make a Fierz transformation, we can write

$$\begin{aligned} 2O_{\pm}^{\overline{c}u} = &\ \overline{c}\gamma^{\mu}(1+\gamma_5)s_{\alpha}\overline{d}_{\beta}\gamma_{\mu}(1+\gamma_5)u_{\beta} \\ &\pm\overline{c}_{\beta}\gamma^{\mu}(1+\gamma_5)s_{\alpha}\overline{d}_{\alpha}\gamma_{\mu}(1+\gamma_5)u_{\beta}, \end{aligned} \tag{9.3.4}$$

which exhibits the symmetry (antisymmetry) of O_+ (O_-) in the color indices of the $\overline{c}$ and $\overline{d}$ fields. Thus, in O_+ the $\overline{c}$ and $\overline{d}$ are combined into a 6 of color $SU(3)$ while in O_- they are in a $\overline{3}$. Thus, in O_+ (O_-), gluon exchange is repulsive (attractive).

Quantitatively, the anomalous dimension of (9.3.2) is proportional to

$$T_a^{\overline{c}}T_a^{\overline{d}} + T_a^{s}T_a^{u}, \tag{9.3.5}$$

where T_a^F are the color $SU(3)$ generators acting on the F field. The operators $O_\pm$ are eigenstates of (9.3.5) with eigenvalues $\frac{2}{3}$ for O_+ and $-\frac{4}{3}$ for O_-. Thus, to order g^2, the operators O_+ and O_- are multiplicatively renormalized.

If we write the effective couplings as

$$\left(h_+(\mu)O_+^{\overline{c}u} + h_-(\mu)O_-^{\overline{c}u}\right)\cos^2\theta_c G_F/\sqrt{2}, \tag{9.3.6}$$

the β functions for h_+ are

$$\begin{aligned} \beta_+ &= \tfrac{1}{4\pi^2}\,g^2h_+ \\ \beta_- &= -\tfrac{1}{2\pi^2}\,g^2h_-. \end{aligned} \tag{9.3.7}$$

(9.3.1) gives the boundary condition $h_+(M_W) = h_-(M_W) = 1$. Then (9.2.20) and appropriate matching conditions (for $m_t < M_W$) gives

$$h_+(m_c) = \left(\frac{\alpha_s(m_c)}{\alpha_s(m_b)}\right)^{-6/25} \left(\frac{\alpha_s(m_b)}{\alpha_s(m_t)}\right)^{-6/23} \left(\frac{\alpha_s(m_t)}{\alpha_s(M_W)}\right)^{-6/21},$$

$$h_-(m_c) = \left(\frac{\alpha_s(m_c)}{\alpha_s(m_b)}\right)^{12/25} \left(\frac{\alpha_s(m_b)}{\alpha_s(m_t)}\right)^{12/23} \left(\frac{\alpha_s(m_t)}{\alpha_s(M_W)}\right)^{12/21}. \tag{9.3.8}$$

For example, for $\Lambda = 100$ MeV, $m_c = 1.4$ GeV, $m_b = 4.6$ GeV, and $m_t = 30$ GeV, this gives

$$h_+(m_c) \simeq 0.8$$

$$h_-(m_c) \simeq 1.5 \tag{9.3.9}$$

Clearly, the QCD corrections enhance h_- compared to h_+. Thus, barring some accident in the matrix elements, we would expect decays produced by the O_- operator to be more important than those produced by the O_+ operator.

The $O_\pm$ operators are distinguished by their internal symmetry properties as well as their color properties. They transform according to different irreducible representations of left–chiral $SU(3)$, O_- as a (2, 0) (the 6) and O_+ as a (1, 2) (the $\overline{15}$). Indeed, things had to work out this way. Since the QCD interactions conserve the internal $SU(3)_L \times SU(3)_R$ symmetry, the operators that are multiplicatively renormalized must transform like irreducible representations. Thus, here and in general, the operators that are enhanced or suppressed by QCD effects will be distinguished by different internal–symmetry properties. In this case, we expect a modest enhancement of the $SU(3)_L$ 6. Because the momenta involved in charm decay are large, the chiral structure of the operator is probably not terribly relevant. But O_+ and O_- also transform as $\overline{15}$ and 6, respectively.

The most striking thing about the O_- operator is that it is a V–spin singlet. V–spin is the $SU(2)$ subgroup of $SU(3)$ under which the u and s quarks form a doublet. The $\bar{d}$ and c are singlets under V–spin, and in the O_- the u and s fields are combined antisymmetrically into a V–spin singlet. Thus, we expect the charm decays that do not change V–spin, so-called $\Delta V = 0$ decays, to be enhanced. The experimental situation is still obscure.

9.4 Penguins and the $\Delta I = \frac{1}{2}$ Rule

Now that we have seen the renormalization group in action in enhancing the $\Delta I = 0$ decays of charmed particles, we can go on to discuss the interesting and more complicated subject of strange–particle decays. At the W scale, the four–fermion coupling that produces the nonleptonic s–quark decay has the form (ignoring mixing to the t and b)

$$\sin\theta_c \cos\theta_c \frac{G_F}{\sqrt{2}} \left(\bar{d}\gamma_\mu(1+\gamma_5)u\bar{u}\gamma_\mu(1+\gamma_5)s\right.$$

$$\left.\bar{d}\gamma^\mu(1+\gamma_5)c\bar{c}\gamma_\mu(1+\gamma_5)s\right). \tag{9.4.1}$$

As we will see, the new feature here that makes it qualitatively different from (9.3.1) is the appearance of quark fields (u, c) along with the corresponding quark field ($\overline{u}$, $\overline{c}$).

Under $SU(3)_L$ and ordinary $SU(3)$, (9.4.1) transforms like a linear combination of a 27, which contains both isospin-$\frac{3}{2}$ and isospin-$\frac{1}{2}$ terms, and an 8, which is pure isospin-$\frac{1}{2}$. Phenomenologically, it is clear that there is a large enhancement of the isospin-$\frac{1}{2}$ decays. Consider, for example, nonleptonic K decay. K^0 decay into 2 π's (as seen in the decay of the K_s, which is a linear superposition of K^0 and $\overline{K}^0$) is much faster than K^+ decay into 2 π's. The two pions in the final state are symmetric in their space wave functions ($l = 0$) and therefore symmetric in isospin. Thus the isospin of the final state is either 0 or 2. But while the neutral final state in K^0 decay can have isospin 0, the charged final state in K^+ decay cannot. Therefore the isospin in the final state in $K^+ \to \pi^+\pi^0$ is 2, and the decay is pure $\Delta I = \frac{3}{2}$. K^0 decay, on the other hand, can be $\Delta I = \frac{1}{2}$, and indeed the ratio of$\pi^0\pi^0$ to $\pi^+\pi^-$ is about $1:2$, as one would expect if the isospin of the final state is 0. Furthermore, the $K^0 \to \pi\pi$ is much faster than any semileptonic decays while the $K^+ \to \pi\pi$ decay is smaller than the two–body leptonic decay. All this suggests that the $\Delta I = \frac{3}{2}$ decays are suppressed while the $\Delta I = \frac{1}{2}$ decays are enhanced by a large factor.

The same pattern appears in the nonleptonic hyperon decays. The $\Delta I = \frac{1}{2}$ amplitudes are consistently larger by at least an order of magnitude. There is also some evidence from the chiral Lagrangian analysis of the nonleptonic decays that it is the $SU(3)$ octet part of the weak Hamiltonian that is enhanced (see Section 6.6). Let us try to understand this enhancement in the effective–field–theory language.

For scales above the charmed quark mass, the analysis is precisely the same as the analysis of charm decay. The reason is that in this region we ignore the charmed quark mass and the effective theory has an $SU(4)$ symmetry (of course, there are larger symmetries above the b and t masses, but these don't concern us because our operators do not involve b and t quarks). Under this $SU(4)$ symmetry both (9.3.1) and (9.4.1) transform as linear combinations of components of 20 and 84 dimensional irreducible representations of $SU(4)$. The operators in $SU(4)$ representation are related by the $SU(4)$ symmetry and so, of course, must be renormalized in the same way. The $O_\pm^{\overline{c}u}$ operators in (9.3.3) are in the 84 and 20, respectively. The corresponding operators here are the symmetric and antisymmetric combinations:

$$\begin{aligned} 2\left(O_\pm^{\overline{u}u} - O_\pm^{\overline{c}c}\right) = \; & \overline{d}\gamma^\mu(1+\gamma_5)u\overline{u}\gamma_\mu(1+\gamma_5)s \\ & \pm\overline{d}\gamma^\mu(1+\gamma_5)s\overline{u}\gamma_\mu(1+\gamma_5)u \\ & -\overline{d}\gamma^\mu(1+\gamma_5)c\overline{c}\gamma_\mu(1+\gamma_5)s \\ & \pm\overline{d}\gamma^\mu(1+\gamma_5)s\overline{c}\gamma_\mu(1+\gamma_5)c. \end{aligned} \tag{9.4.2}$$

The coefficients of these operators satisfy the same renormalization group equations as the h_+ in Section 9.3. Thus, the coefficient of)$_-$ at $\mu = m_c$ is enhanced while the coefficient of O_+ is suppressed. Notice that under $SU(3)_L$, O_+ is a linear combination of 27 and 8 while O_- is pure 8. Thus, while this mechanism already enhances the octet contribution, but not by an enormous factor. Something more is required to explain the observed dominance of $\Delta I = \frac{1}{2}$ decays.

For scales below the charmed quark mass, the c quark is removed from the effective theory, thus the structure of the effective $\Delta S = 1$ Hamiltonian changes. In lowest order, the matching conditions are that the coefficients of operators that do not involve c or $\overline{c}$ are continuous at $\mu = m_c$ while operators containing c or $\overline{c}$ fields are just dropped, their coefficients set to zero at $\mu = m_c$. Thus, for $\mu = m_c$ in the effective theory below m_c, the $\Delta S = 1$ effective Hamiltonian is

$$\left(h_+(m_c)O_+^{\overline{u}u} + h_-(m_c)O_-^{\overline{u}u}\right)\sin\theta_c\cos\theta_c G_F/\sqrt{2}, \tag{9.4.3}$$

where the $h_\pm(m_c)$ are given by (9.3.9). Above m_c, (9.4.2) are the only dimension–6 operators transforming like 84 and 20 of $SU(4)_L$ with appropriate quantum numbers. Thus, they must be multiplicatively renormalized. The situation for the (9.4.3) in the effective theory below m_c is more complicated. The $O_+^{\overline{u}u}$ operator transforms like a linear combination of 27 and 8 under $SU(3)_L$ while $O_-^{\overline{u}u}$ is pure 8. The 27 part is multiplicatively renormalized, but the 8's can mix with a variety of other operators that transform similarly under $SU(3)_L$. Any dimension–6, $SU(3)_L$–octet (and $SU(3)_R$ singlet), color–singlet operator is a linear combination of the operators

$$\begin{aligned}
O_1 &= \overline{d}\gamma^\mu(1+\gamma_5)s\overline{q}\gamma_\mu(1+\gamma_5)q && (9.4.4)\\
O_2 &= \overline{d}\gamma^\mu(1+\gamma_5)s\overline{q}\gamma_\mu(1-\gamma_5)q && (9.4.5)\\
O_3 &= \overline{d}T_a\gamma^\mu(1+\gamma_5)s\overline{q}T_a\gamma_\mu(1+\gamma_5)q && (9.4.6)\\
O_4 &= \overline{d}T_a\gamma^\mu(1+\gamma_5)s\overline{q}T_a\gamma_\mu(1-\gamma_5)q && (9.4.7)\\
O_5 &= \frac{1}{g}\overline{d}T_a\gamma^\mu(1+\gamma_5)s[D^\nu G_{\mu\nu}]_a, && (9.4.8)
\end{aligned}$$

where $\overline{q}q$ means the $SU(3)$ invariant sum $\overline{u}u + \overline{d}d + \overline{s}s$, T_a are the color $SU(3)$ charges, and $G^{\mu\nu}$ is the gluon–field strength. (9.4.4) and (9.4.6) have nonzero coefficients at $\mu = m_c$. But nonzero coefficients for all of these are induced by renormalization. Note that the $1/g$ is included in (9.4.8) because (9.4.6), (9.4.7), and (9.4.8) are related by the equation of motion

$$(D_\mu G^{\mu\nu})_a = g\overline{q}T_a\gamma^\nu q. \tag{9.4.9}$$

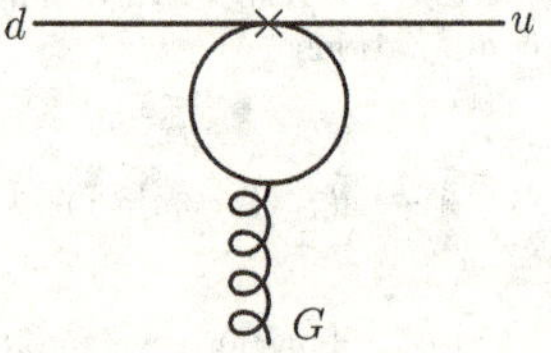

Figure 9-1:

We will not discuss the β function in detail here (but see Problem 9–4). The important thing is that all terms (9.4.4)–(9.4.8) are produced at low momenta. Thus, there are many sources of $\Delta I = \frac{1}{2}$, $SU(3)_L$ octet decay. For example, O_5 is produced from O_3 by the diagram shown in figure 9-1. If we hook the gluon in (Figure 9–1) onto a quark line and take the O_3 vertex apart to indicate that O_3 is produced by W exchange, this becomes the so-called "Penguin" diagram (Figure 9–2). This may be interesting as an example of capricious use of jargon, but what does it have to do with the $\Delta I = \frac{1}{2}$? The hope is that the matrix elements of O_2, O_4, and O_5 that involve gluon fields or both the left-handed and right-handed quark fields are much larger than the corresponding matrix elements of O_1 and O_3 that involve only left-handed quarks. There is no very good theoretical reason to believe that this is true. But it is not unreasonable. The basis for the speculation is called the "vacuum insertion approximation". Consider, for example, the two operators

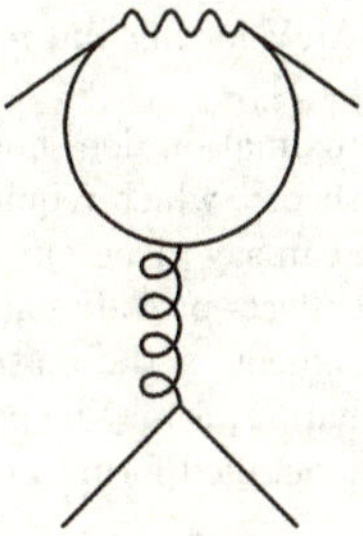

Figure 9-2:

$$\overline{d}\gamma^\mu(1+\gamma_5)u\overline{u}\gamma_\mu(1+\gamma_5)s$$
$$\overline{d}(1-\gamma_5)u\overline{u}(1+\gamma_5)s. \tag{9.4.10}$$

These are contained (in Fierzed form) in O_3 and O_4, respectively. Consider the matrix elements between a K^- and π^- of these operators. In the vacuum insertion approximation, one inserts a complete set of states in the middle of the operators (9.4.10). This is already a peculiar thing to do, since the renormalized operator cannot really be considered as a product of two factors. But then in the sum over states, only the vacuum state is kept. This is an even more peculiar thing to do, but at least it makes the calculation easy:

$$\begin{gathered}\langle\pi^-|\overline{d}\gamma^\mu(1+\gamma_5)u\overline{u}\gamma_\mu(1+\gamma_5)s|K^-\rangle \\ \simeq \langle\pi^-|\overline{d}\gamma^\mu(1+\gamma_5)u|0\rangle\langle 0|\overline{u}\gamma_\mu(1+\gamma_5)s|K^-\rangle \\ = f_\pi f_K p_\pi^\mu p_{k\mu}\end{gathered} \tag{9.4.11}$$

$$\begin{gathered}\langle\pi^-|\overline{d}(1-\gamma_5)u\overline{u}(1+\gamma_5)s|K^-\rangle \simeq \langle\pi^-|\overline{d}(1-\gamma_5)u|0\rangle\langle 0|\overline{u}(1+\gamma_5)s|K^-\rangle \\ = -\langle\pi^-|\overline{d}\gamma_5 u|0\rangle\langle 0|\overline{u}\gamma_5 s|K^-\rangle \\ = -\frac{1}{(m_d+m_u)}\cdot\frac{1}{(m_u+m_s)} \\ \langle\pi^-|\partial_\mu\overline{d}\gamma_5\gamma^\mu u|0\rangle\cdot\langle 0|\partial_\nu\overline{u}\gamma_5\gamma^\nu s|K^-\rangle \\ = -f_\pi f_K\frac{m_\pi^2}{(m_d+m_u)}\cdot\frac{m_K^2}{(m_u+m_s)}.\end{gathered} \tag{9.4.12}$$

If the quark mass are small, say $\sim$ 10 MeV for m_d and m_u and $\sim$ 200 MeV for m_s, (9.4.12) can be significantly enhanced compared to (9.4.11),

Clearly, the vacuum insertion approximation does not make a great deal of sense. In fact, it is inconsistent with chiral perturbation theory, which requires that the matrix element vanish as the quark masses go to zero. The best we can say given our present level of theoretical understanding is that short–distance renormalization effects provide some enhancement of the $\Delta I = \frac{1}{2}$ amplitude. The rest must come from the enhancement of the matrix elements of one or more of the octet operators. As we will see in the next chapter, it may be possible to test the idea that the "Penguin" operator is the culprit by studying the detailed form of CP violation in the K mesons.

Problems

9-1. Derive (9.3.7). Work in Landau gauge.

9-2. Write (9.4.1) as a linear combination of terms which transform like a component of an $SU(3)$ 27 with $I = 3/2$, a component of a 27 with $I = 1/2$, and a component of an 8.

9-3. Below m_c, the $\Delta S = 1$ Hamiltonian has the form $h_x(\mu)\, O_x$, where the O_x are the operators in (9.4.4-9.4.8). The β functions for the h_x in lowest order have the form

$$\beta_x = g^2 A_{xy} h_y$$

Find all the components of A_{xy}.

9-4. Find $h_\pm(\mu)$ such that $\alpha_s(\mu) = 1$.

10 — The Neutral K Mesons and CP Violation

10.1 $K^0 - \overline{K}^0$ Mixing

We have postponed until now a detailed discussion of the neutral K mesons because we will need most of the theoretical ideas we have discussed in the previous chapters to understand it. Indeed, the study of the neutral K–meson system played a crucial role in the development of many of these ideas.

The most striking feature of the neutral K mesons is the mixing of the K^0 and $\overline{K}^0$. The K^0 mesons with strangeness $S = 1$ and $\overline{K}^0$ with $S = -1$ are the states that are produced by the strangeness–conserving QCD strong interactions. They are antiparticles of one another and therefore have identical mass terms because of CPT invariance. If strangeness were exactly conserved, there would be no interactions that could cause mixing between K^0 and $\overline{K}^0$. The K^0 and $\overline{K}^0$ would be exactly degenerate mass eigenstates like the electron and positron. States that are linear combinations of K^0 and $\overline{K}^0$ would be forbidden by superselection rules, which simply means that there would be no reason to consider such states. But the weak interactions do not conserve strangeness. Thus, they can cause mixing between K^0 and $\overline{K}^0$. Of course, they also cause the K^0 and $\overline{K}^0$ to decay. As we will see in detail, the amusing thing about this system is that the neutral K mesons mix as fast as they decay.

In this section and the next, we will discuss K^0-$\overline{K}^0$ mixing, ignoring the Kobayashi-Maskawa mixing of the third family. In this theory, CP is conserved, and the mass eigenstates behave simply under CP. In the most natural basis, K^0 and $\overline{K}^0$ are charged conjugates, thus

$$C|K^0\rangle = |\overline{K}^0\rangle, \quad C|\overline{K}^0\rangle = |K^0\rangle. \tag{10.1.1}$$

Then, because the K's are pseudoscalar, the states with zero three–momentum satisfy

$$CP|K^0\rangle = -|\overline{K}^0\rangle, \quad CP|\overline{K}^0\rangle = |K^0\rangle. \tag{10.1.2}$$

The mass eigenstates are therefore the CP even and odd states

$$|K_1\rangle = \frac{|K^0\rangle - |\overline{K}^0\rangle}{\sqrt{2}}, \quad |K_2\rangle = \frac{|K^0\rangle - \sqrt{K^0}\rangle}{\sqrt{2}}. \tag{10.1.3}$$

These states have very different properties because their nonleptonic decay modes are radically different. Two pions, $\pi^0\pi^0$ or $\pi^+\pi^-$, in an $l = 0$ state must be CP even. Thus, if CP is conserved, only K_1 can decay into two π's. In the simplified world we are considering, with no mixing of the third family, this statement is exactly true. In the real world, CP violation is a very small effect, thus, the picture that emerges in the limit of CP invariance will look very much like the real world.

The importance of the two–pion final states is simply that there is a much larger phase space for it than for the three–pion final state. Thus, K_1, which can decay into 2π's, decays much faster than K_2, which cannot. The two mass eigenstates have very different lifetimes. This, of course, is just what is observed. The K_s (S for short) and K_L (L for long) are linear combinations of K^0 and $\overline{K}^0$ with very different lifetimes. A beam of neutral K's produces a splash of two–pion decays and a trickle of three–pion decays farther downstream, where essentially only the K_L component survives.

When a K_L beam passes through matter, the K^0 and the $\overline{K}^0$ components behave differently. This regenerates a K_s component of the beam, producing another splash of 2π's. This phenomenon can be used to make an accurate measurement of the mass difference between the K_s and K_L. If two regenerators (pieces of matter) are placed in a K_L beam some distance apart, the K_s component produced by the first regenerator interferes with the K_s component produced by the second. When the distance between the regenerators is varied, the K_s amplitude and thus the 2π yield oscillates with a wave number proportional to the mass difference between K_L and K_s, The result of such experiments is

$$\begin{aligned}\Delta m &= m_{K_L} - m_{K_s} = 0.535 \pm .002 \times 10^{10}\text{sec}^{-1} \\ &= 3.52 \times 10^{-6}\text{eV}.\end{aligned} \tag{10.1.4}$$

It is the fact that the mass difference is so small that makes regeneration experiments interesting. The quantum–mechanical interference effects are spread out over macroscopic distances.

Theoretically, the smallness of Δm is equally interesting. It played an important role in the prediction of the charmed quark. Today, it provides one of the most important constraints on models in which the Higgs doublet of the standard $SU(2) \times U(1)$ model is replaced by a dynamical symmetry–breaking mechanism and on other variants of the standard model.

It is useful to describe the mixing and decay of a neutral K meson in terms of a two–state wave function

$$|\psi(t)\rangle = A(t)|K^0\rangle + B|\overline{K}^0\rangle, \tag{10.1.5}$$

where $\psi(t)$ is a column vector

$$\psi(t) = \begin{pmatrix} A(t) \\ B(t) \end{pmatrix} \tag{10.1.6}$$

$A(t)$ $B(t)$ is the amplitude for finding the system as a $K^0 (\overline{K}^0)$. The time development of the system can now be described by a matrix Hamiltonian:

$$i\frac{d}{dt}\psi(t) = H\psi(t) \tag{10.1.7}$$

$$H = \begin{pmatrix} M - i\Gamma/2 & M_{12} - i\Gamma_{12}/2 \\ M_{12}^* - i\Gamma_{12}^*/2 & M - i\Gamma/2 \end{pmatrix} \tag{10.1.8}$$

where M and Γ are real. The form of (10.1.8) may require some explanation. The anti-Hermitian part of H describes the exponential decay of the K–meson system due to weak interaction processes. Because the weak decay amplitudes are proportional to G_F, all the Γ's are of order G_F^2. The Hermitian part is called a mass matrix. If all the Γ's were zero, the system would evolve without decay, but the off-diagonal elements in the mass matrix would still cause mixing of K^0 with $\overline{K}^0$. M is just the K^0 mass. M_{12}, however, is a $\Delta S = 2$ effect. It must be of order G_F^2, because the weak interactions in lowest order change strangeness only by ± 1 (once the GIM mechanism has been incorporated). If this were not the case, the mass difference Δm (which we will see below is $\sim 2M_{12}$) would be much larger than what is observed.

In (10.1.8), we have assumed CPT invariance by taking the two diagonal elements to be equal. If we also, for the moment, impose CP invariance, we must take

$$M_{12} = M_{12}^*, \quad \Gamma_{12} = \Gamma_{12}^*, \quad CP. \tag{10.1.9}$$

Then the K_1 and K_2 states correspond to wave functions

$$\psi_1(\tau) = \frac{1}{\sqrt{2}}\begin{pmatrix} a(t) \\ -a(t) \end{pmatrix}, \quad \psi_2(\tau) = \frac{1}{\sqrt{2}}\begin{pmatrix} a(t) \\ a(t) \end{pmatrix}. \tag{10.1.10}$$

They have masses $M \mp M_{12}$ and decay rates $\Gamma \mp \Gamma_{12}$. Thus, the fact that the K_1 decays much faster than the K_2 implies

$$\Gamma_{12} \simeq -\Gamma. \tag{10.1.11}$$

10.2 The Box Diagram and the QCD Corrections)

We would like to calculate the parameters in (10.1.8) from the underlying $SU(3) \times SU(2) \times U(1)$ model of strong, weak, and electromagnetic interactions. In this, as in so many other applications, we will be frustrated by our inability to deal with the low–energy strong interactions. But as usual, we can gather most of our ignorance into a few strong interaction matrix elements. The new feature here, compared with the analysis of Chapter 9, is that we must work to second order in the weak interactions. The only parameter in (10.1.8) in which this new feature appears is M_{12}.

Let us write the Hamiltonian of the world is a power series in G_F,

$$H = H_0 + H_1 + H_2 + \cdots \tag{10.2.1}$$

where H_j contains all the interactions proportioned to G_F^j. The $|K^0\rangle$ and $|\overline{K}^0\rangle$ states are eigenstates of H_0 because they are degenerate, stable particles when the weak interactions are turned off. For zero–momentum states normalized to 1, perturbation theory gives

$$M_{12} = \langle K^0|H_2|\overline{K}^0\rangle + \sum_n \frac{\langle K^0|H_1|n\rangle\langle n|H_1|\overline{K}_0\rangle}{m_{K^0} - E_n}. \tag{10.2.2}$$

The leading contributions are both second order in G_F because H_1 only changes strangeness by ± 1. In terms of states with the conventional relativistically invariant–continuum normalization, the first term is

$$M_{12} \simeq \frac{1}{2m_K}\langle K^0|\mathcal{H}_2(0)|\overline{K}^0\rangle + \cdots \tag{10.2.3}$$

where $\mathcal{H}_2(0)$ is the second–order weak Hamiltonian density. We should be able to get some idea of magnitude of this term by using the effective–field–theory technology and estimating the matrix element. The second term in (10.2.2), however, depends even more sensitively on the details of low–energy strong interactions because it involves a sum over all the low–lying $S = 0$ mesonic states. Thus, even if we manage a reliable estimate of (10.2.3), we will still not know M_{12} very well. Nevertheless, this estimate is interesting and important, as we will see.

At the scale M_W and just below, there is no $\Delta S = 2$ term in the effective Lagrangian in order G_F^2 because of the GIM mechanism. To see how this works in detail, consider the box diagram in Figure 10–1. The important point is that for large loop momentum, the contributions of the u and c quarks cancel on each quark line. This is a consequence of the GIM mechanism. All

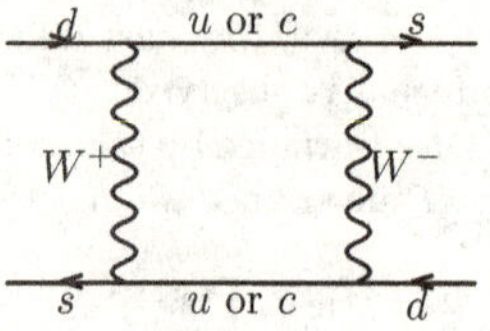

Figure 10-1:

strangeness–changing interactions would disappear if the charge-$\frac{2}{3}$ quarks were degenerate. Thus, each strangeness–changing interaction, each quark line in Figure 10–1, is proportional to $m_c^2 - m_u^2$ and the diagram is quadratically convergent. Because of the extra convergence produced by the GIM mechanism, (10.2.4) receives no order G_F^2 contribution from momenta much larger than m_c. For $m_t < \mu < M_W$, the pieces of H_1 get renormalized by the QCD interactions as discussed in Chapter 9.

In the simplified theory with no mixing to the third family, nothing very exciting happens at $\mu \simeq m_t$. As μ drops below the t and b quark thresholds, the β functions change, but no $\Delta S = 2$ terms are introduced.

At $\mu \simeq m_c$, a $\Delta S = 2$ term is induced. At this point, the c quark is removed from the theory, and the GIM mechanism no longer operates. Thus, the box diagram and its QCD corrections induce a $\Delta S = 2$ term in the effective Lagrangian at $\mu \simeq m_c$. In other words, the $\Delta S = 2$ term is produced in the matching condition between the effective theories above and below the charmed–quark mass. In the first, the $\Delta S = 1$ Hamiltonian involves the operators $O_\pm^{\overline{u}u} - O_\pm^{\overline{c}c}$ [see (9.4.2)] and the corresponding $\Delta C = \pm 1$ operators $O_\pm^{\overline{c}u}$ and $O_\pm^{\overline{u}c}$ [see (9.3.2)]. The combinations

$$\sin\theta_c \cos\theta_c (O_\pm^{\overline{u}u} - O_\pm^{\overline{c}c}) + \cos^2\theta_c O_\pm^{\overline{c}u} - \sin^2\theta_c O_\pm^{uc} \tag{10.2.4}$$

retain the GIM structure. But in the effective theory below m_c, the relevant $\Delta S = 1$ operators are $\sin\theta_c \cos\theta_c O_\pm^{uu}$, from which terms involving c quark fields have been expanded. In a second–order calculation, the $\Delta S = 2$ effect produced by squaring (10.2.4) is finite and proportional to $M_c^2 - m_u^2$ because of GIM, but that produced by two $O_\pm^{\overline{u}u}$ operators has no term proportional to m_c^2 (after a quadratic divergence is removed by dimensional regularization). The $\Delta S = 2$ term in the effective theory below m must be present to reproduce the finite result in the theory above m_c. A simple generalization of the box diagram calculation yields the result of the $\Delta S = 2$ Hamiltonian density renormalized at $\mu < m_c$ (ignoring terms proportional to m_u^2),

$$\begin{aligned} \mathcal{H} = \; & \sin^2\theta_c \cos^2\theta_c m_c^2 \frac{G_F^2}{16\pi^2}\eta_1 \\ & \overline{d}\gamma^\mu(1+\gamma_5)s\overline{d}\gamma_\mu(1+\gamma_5)s, \end{aligned} \tag{10.2.5}$$

where

$$\eta_1 = \left[\frac{\alpha_s(m_c)}{\alpha_s(\mu)}\right]^{6/27} \cdot \left[\frac{3}{2}h_+(m_c)^2 - h_+(m_c)h_-(m_c) + \frac{1}{2}h_-(m_c)^2\right]. \tag{10.2.6}$$

Note that we have included in η_1 a factor that incorporates the renormalization down to a scale μ of the $\Delta S = 2$ operator in the theory below m_c. The operator is actually $O_+^{\overline{d}s}$. Here, as in the undressed box diagram, the $\Delta S = 2$ effect is proportional to m_c^2 (renormalized at $\mu \sim m_c$). It was this feature that led Lee and Gaillard to suggest in 1974 that the charmed quark and the hadron states built out of it had to be relatively light. Otherwise, the contribution to M_{12} from (10.2.5) would be too large. Such a statement requires some estimate of the matrix element of the $\Delta S = 2$ operator. Gaillard and Lee used a version of the vacuum insertion approximation [see (9.4.11) and (9.4.12)]. If we insert a vacuum state between all possible pairs of quark fields in $O_+^{\overline{d}s}$, we get

$$\langle K^0 | O_+^{\overline{d}s} | \overline{K^0} \rangle \simeq \frac{8}{3} f_K^2 m_K^2. \tag{10.2.7}$$

To evaluate Δm, we need to know η_1 and thus decide what value of the renormalization scale μ to take. Of course, if we really knew how to calculate the matrix elements, it would not matter because the matrix element of $O_+^{\overline{d}s}$ renormalized at μ would have an explicit μ dependence that would cancel the μ dependence of η_1. Here, all we can do is to guess what value of μ is most likely to make (10.2.7) approximately correct. Presumably, we should choose some small μ such that $\alpha_s(\mu) \simeq 1$. To be definite (and for ease of comparison with the papers of Gilman and Wise), we choose μ such that

$$\alpha(\mu) = 1.$$

Then

$$\eta_a \simeq 0.7 \tag{10.2.8}$$

Combining (10.2.5), (10.2.7), and (10.2.8) gives a contribution to Δm,

$$\sim (m_c/1.4\,\text{GeV})^2 \times 1.7 \times 10^{-6}\,\text{eV}. \tag{10.2.9}$$

This led Gaillard and Lee and others to believe that m_c was less than a few GeV, well before there was any clear experimental indication of the existence of charm.

In fact, it is possible to do a better job of estimating the matrix elements by using the measured value of the $\Delta I = \frac{3}{2}$ decay rate of $K^+ \to \pi^+\pi^0$. The argument goes as follows. The $\Delta I = \frac{3}{2}$, $\Delta S = 1$ part of the weak Hamiltonian density is

$$\frac{h_+(\mu)}{3}(2O_+^{\overline{u}u} - O_+^{\overline{d}d}) \sin\theta_c \cos\theta_c G_F/\sqrt{2}. \tag{10.2.10}$$

The operator $2O_+^{\overline{u}u} - O_+^{\overline{d}d}$ is an $SU(3)_L$ 27, like $O_+^{\overline{d}s}$. Both have the form

$$T_{lm}^{jk} \overline{\psi^j} \gamma^\mu (1+\gamma_5) \psi^l \overline{\psi^k} \gamma_\mu (1+\gamma_5) \psi^m, \tag{10.2.11}$$

where T is a traceless tensor, symmetric in upper and lower indices. For $2O_+^{\overline{u}u} - O_+^{\overline{d}d}$ the nonzero components of T are

$$\begin{gathered} T_{13}^{12} = T_{13}^{21} = T_{31}^{12} = T_{31}^{21} = \tfrac{1}{2} \\ T_{23}^{22} = T_{32}^{22} = -\tfrac{1}{2}. \end{gathered} \tag{10.2.12}$$

For $O_+^{\overline{d}s}$, the nonzero component is

$$T_{33}^{22} = 1. \tag{10.2.13}$$

From the measured $K^+ \to \pi^+\pi^0$ width, we can determine the magnitude of the matrix element

$$|\langle \pi^+\pi^0|2O_+^{\overline{d}d}|K^+\rangle| \simeq \frac{3\times 10^{-2}\text{GeV}^3}{h_+(\mu)}. \tag{10.2.14}$$

This estimate includes a small correction from $\eta_1 - \pi$ mixing.

To determine the matrix element of $O_+^{\overline{d}s}$, we compare the two matrix elements in an effective chiral theory of the Goldstone bosons at low energies. Here the operator (10.2.11) can be represented by a sum over all operators with the same $SU(3)_L \times SU(3)_R$ properties with coefficients that are unknown (although in principle they could be determined by nonperturbative matching conditions). But for the $SU(3)_L$ 27, there is a unique operator in the effective theory that involves only two derivatives or one power of μM. Thus, all the matrix elements of (10.2.11) are determined in terms of a single parameter. Explicitly, (10.2.11) becomes

$$a T^{jk}_{lm} (\Sigma\partial_\mu\Sigma^\dagger)^l_j (\Sigma\partial^\mu\Sigma^\dagger)^m_k. \tag{10.2.15}$$

Then

$$|\langle \pi^+\pi^0|2O_+^{\overline{u}u} - O_+^{\overline{d}d}|K^+\rangle| = 6|a|\sqrt{2}\,\frac{m_K^2 - m_\pi^2}{f_\pi^3} \tag{10.2.16}$$

and

$$\langle K^0|O_+^{\overline{d}s}|\overline{K^0}\rangle = 8am_K^2/f_\pi^2. \tag{10.2.17}$$

Note that only f_π appears, not f_K, because we are working to lowest order in symmetry breaking and our chiral theory doesn't know the difference. Comparing (10.2.14), (10.2.16), and (10.2.17), we find

$$|\langle K^0|O_+^{\overline{d}s}|\overline{K^0}\rangle| \simeq \frac{4\times 10^{-3}\,\text{GeV}^4}{h_+(\mu)}. \tag{10.2.18}$$

Note that this argument does not determine the sign of the matrix elements. But it does have the appropriate dependence on μ because it was determined consistently from a physical quantity. For μ such that $\alpha_s(\mu) = 1$, (10.2.18) is smaller than the vacuum insertion result by a factor of more than 2.

With this estimate for the matrix element, (10.2.5) gives only a small fraction of the measured value of Δm. The mixing to the third family, which we have ignored so far, can give an additional contribution. The rest must come from the second term in (10.2.2).

10.3 The Gilman-Wise $\Delta S = 2$ Hamiltonian

[1] When mixing with the third family is turned on, the discussion of the previous section must be generalized. The general case has been studied by Gilman and Wise in a different language. In this section, I explain and interpret their results in the effective–field–theory language.

In the standard six–quark model, the $\Delta S = 1$ Hamiltonian density just below M_W is

$$\begin{aligned}\mathcal{H} &= \tfrac{G_F}{\sqrt{2}} \textstyle\sum_{x,y} U_s^x U_y^{\dagger d}\, \overline{x}\gamma^\mu(1+\gamma_5)s\overline{d}\gamma_\mu(1+\gamma_5)y \\ &= \tfrac{G_F}{\sqrt{2}} \textstyle\sum_{x,y} U_s^x U_y^{\dagger d}(O_+^{\overline{x}y} + O_-^{\overline{x}y}),\end{aligned} \tag{10.3.1}$$

[1]F. Gilman and M. B. Wise, *Phys. Rev.* **D27**:1128–1141, 1983.

where the sums run over the charge-$\frac{2}{3}$ quark fields, u, c, and t and where U is the KM matrix (3.5.32). The $\Delta S = 2$ Hamiltonian density produced to second order in U has the form

$$\mathcal{H} = \xi_c^2 \mathcal{H}_c + \xi_t^2 \mathcal{H}_t + 2\xi_c \xi_t \mathcal{H}_{ct}, \tag{10.3.2}$$

where

$$\begin{aligned} \xi_c &= U_s^c U_c^{\dagger d} = -s_1 c_2 \left(c_1 c_2 c_3 + s_2 s_3^{e^{i\delta}}\right) \\ \xi_t &= U_s^t U_t^{\dagger d} = -s_1 s_2 \left(c_1 c_2 c_3 - c_2 s_3^{e^{i\delta}}\right). \end{aligned} \tag{10.3.3}$$

The striking fact that the flavor structure of $\mathcal{H}_2$ depends only on the two parameters ξ_c and ξ_t can be understood as follows. The terms in $\mathcal{H}_1$ with $x = y$ have $\Delta S = 1$ but do not change u, c, or t number; thus, they can mix with each other and with other $\Delta S = 1$ terms due to QCD interactions. But they depend only on ξ_c and ξ_t because the unitarity of U implies

$$U_s^u U_u^{\dagger d} = -\xi_c - \xi_t. \tag{10.3.4}$$

The terms with $x \neq y$ carry nonzero, u, c, and t number. Thus, to produce a $\Delta S = 2$ term with zero, u, c, and t number, $O_{\pm}^{\overline{x}y}$ must appear together with $O_{\pm}^{\overline{y}x}$, so the $\Delta S = 2$ effect depends on

$$(U_s^x U_y^{\dagger d})(U_s^y U_x^{\dagger d}) = (U_s^x U_x^{\dagger d})(U_s^y U_y^{\dagger d}), \tag{10.3.5}$$

which in turn, depends only on ξ_c and ξ_t by virtue of (10.3.3) and (10.3.4).

$\mathcal{H}_c$ is the term that we have already calculated in the previous section, since it is all that survives when s_2 and s_3 go to zero. In the limit $s_2 = s_3 = 0$, $s + 1 = \sin\theta)c$, $\xi_c = -\sin_c \cos theta_c$, and $\xi_t = 0$, and comparing (10.2.5) and (10.3.3) we can write

$$\mathcal{H} = m_c^2 \frac{G_F}{16\pi^2} \eta_1 O_+^{\overline{d}s}, \tag{10.3.6}$$

where η_1 is still given by (10.2.6).

$\mathcal{H}_t$ is a similar term. It would be the only term present if c and b formed a doublet that did not mix with u, d, s, and t. When the t quark is removed at $\mu \simeq m_t$, a $\Delta S = 2$ term is produced that is analogous to (10.2.5), but proportional to m_t^2. Below m_t, the induced $\Delta S = 2$ is renormalized by the QCD interactions. In addition, the remaining $\Delta S = 2$ terms, but because there are no t–quark fields left, none are proportional to m_t^2. Presumably, in the real world, ξ_t is much smaller than ξ_c. The only reason that the $\mathcal{H}_t$ term is important at all is that there is a term proportional to m_t^2. Thus, we can safely neglect the terms in $\mathcal{H}_t$ that are not proportional to m_t^2. The we can write

$$\mathcal{H}_t = m_t^2 \frac{G_F}{16\pi^2} \eta_2 O_+^{\overline{d}s}, \tag{10.3.7}$$

where

$$\begin{aligned} \eta_2 = \quad & \left[\frac{\alpha_s(m_c)}{\alpha_s(\mu)}\right]^{6/27} \cdot \left[\frac{\alpha_s(m_b)}{\alpha_s(m_c}\right]^{6/25} \cdot \left[\frac{\alpha_s(m_t)}{\alpha_s(m_b}\right]^{6/23} \\ & \cdot \left[\tfrac{3}{2} h_+(m_t)^2 - h_+(m_t) h_-(m_t) + \tfrac{1}{2} h_-(m_t)^2\right]. \end{aligned} \tag{10.3.8}$$

For $\alpha_s = 1, ; \eta_2 \simeq 0.6$.

The $\mathcal{H}_{ct}$ term is very different from ξ_c and ξ_t. If we ignore QCD corrections, it comes from a box diagram with one internal c and one t (plus GIM related graphs), and it is

$$\mathcal{H}_{ct} \simeq m_c^2 \ln(m_t^2/m_c^2) \frac{G_F^2}{16\pi^2} O_+^{\bar{d}s}. \tag{10.3.9}$$

The logarithm and factor of m_c^2 (rather than m_t^2) indicate that this term is not produced by the matching condition at $\mu = m_t$, but rather it evolves due to the form of the β functions for $m_t > \mu > m_c$. We expect QCD corrections to modify (10.3.9), so we write

$$\mathcal{H}_{ct} = m_c^2 \ln(m_t^2/m_c^2) \frac{G_F^2}{16\pi^2} \eta_3 O_+^{\bar{d}s}. \tag{10.3.10}$$

Numerical evaluation of η_s (by Gilman and Wise) gives $\eta_3 \simeq 0.4$ for $\alpha_s(\mu) = 1$.

To get an idea of how the QCD corrections work for such a term, define the coupling constant h_{ct} that is the coefficient of $O_+^{\bar{d}s}$ in $\mathcal{H}_{ct}$ at the scale μ. Note that this can be separated uniquely from the coefficients in ξ_c and ξ_t, even though the operator structure is the same because its ξ_c and ξ_t dependence is different. Explicit calculation of the one–loop $\overline{MS}$ matching condition at $\mu \sim m_t$ yields $h_{ct}(\mu) \propto \ln(\mu/m_t)$, and thus

$$h_{ct}(m_t) = 0, \tag{10.3.11}$$

as expected. The renormalization group equation for h_{ct} in one loop has contributions not only from QCD, but also from second–order weak interactions

$$\mu \frac{d}{d\mu} = \beta_{ct} = \frac{g^2}{4\pi^2} h_{ct} + \sum_{j,k} \lambda_{jk} m_c^2 h_j h_k, \tag{10.3.12}$$

where h_j are the relevant $\Delta S = 1$ couplings, m_c^2 is also treated as a μ–dependent coupling, and λ_{jk} are constants that are nonzero in the region $\mu > m_c$. Notice that if QCD is turned off, the first term in β_{ct} is absent and the second is constant so that (10.3.12) integrates trivially to a logarithm. This is the origin (in the effective–field–theory language) of the $\ln(m_t^2/M_c^2)$ in (10.3.9).

With QCD turned on, part of (10.3.12) is intractable. The relevant couplings are the coefficients of the operators $O_\pm^{\bar{c}u}$, $O_\pm^{\bar{u}c}$, $O_\pm^{\bar{u}u} - O_\pm^{\bar{c}c}$, $O_\pm^{\bar{u}u}$, and the operators O_1 through O_4 of (9.4.4)–(9.4.8). The last six mix with one another and make a complete analytic integration of (10.3.12) impossible. But their contributions to h_{ct} is small because the combinations that appear in (10.3.12) are not present at $\mu = m_t$. They are induced only by mixing. The remaining contribution can be integrated analytically. It depends only on m_c^2 and the coefficients of $O_\pm^{\bar{c}u}$, $O_\pm^{\bar{u}c}$, which are proportional to h_{ct} [of (9.3.6)]. We find for this contribution

$$\beta_{ct} = -\frac{G_F^2}{8\pi^2} m_c^2 \left[\frac{3}{2} h_+^2 + \frac{1}{2} h_-^2 - h_+ h_- \right] + \frac{g^2}{4\pi^2} h_{ct}. \tag{10.3.13}$$

Integration of (10.3.13), using the explicit forms of $h_\pm(\mu)$ and $m_c^2(\mu)$, yields the analytic expression found by Gilman and Wise.

10.4 CP Violation and the Parameter ϵ

In general, in the presence of CP violation in the weak interactions, M_{12} and Γ_{12} in (10.1.8) cannot be made simultaneously real. By changing the relative phases of the K^0 and $\overline{K}^0$ states, which amounts to making a diagonal unitary transformation of (10.1.8), we can change the overall phase of M_{12} and Γ_{12}, but we cannot change the relative phase. In the real world, the relative phase is nonzero, that is, $M_{12}\Gamma_{12}^*$ is complex, and CP violation shows up in the neutral K decays.

The eigenstates of H are not ψ_1 and ψ_2, but

$$\psi_{L,\,S} = \frac{1}{\sqrt{2(1+|\epsilon|^2)}} \begin{pmatrix} (1+\epsilon) \\ \pm(1-\epsilon) \end{pmatrix}, \tag{10.4.1}$$

where ϵ is the so-called CP impurity parameter. The corresponding eigenvalues are

$$M_{K_{L,\,S}} - \frac{i}{2}\Gamma_{K_{L,\,S}} = M - \frac{i}{2}\Gamma \pm \frac{\Delta M - i\Delta\Gamma/2}{2}. \tag{10.4.2}$$

The parameters ϵ and $\Delta M - i\Delta\Gamma/2$ satisfy

$$\frac{(1+\epsilon)}{(1-\epsilon)} = 2\cdot\frac{M_{12} - i\Gamma_{12}/2}{\Delta M - i\Delta\Gamma/2} = \frac{1}{2}\frac{\Delta M - i\Delta\Gamma/2}{M_{12}^* - i\Gamma_{12}^*/2}. \tag{10.4.3}$$

Note that (10.4.1) and (10.4.3) depend on CPT invariance, which also makes H very easy to diagonalize because the diagonal terms are equal. Note that the observable quantities ΔM and $\Delta\Gamma$ depend only on the relative phase of M_{12} and Γ_{12}, as they should. In fact,

$$\Delta M^2 - \Delta\Gamma^2/4 = 4|M_{12}|^2 - |\Gamma_{12}|^2, \quad \Delta M\Delta\Gamma = 4\mathrm{Re}(M_{12}\Gamma_{12}^*). \tag{10.4.4}$$

On the other hand, the parameter ϵ does depend on the choice of phase. The standard convention is to choose the phase of the K^0 and $\overline{K}^0$ states to remove the phases from their $\Delta I = \frac{1}{2}$ decay amplitudes except for the effect of the final–state interactions between the pions. Thus

$$\langle\pi\pi(I=0)|H_W|K^0\rangle = -\langle\pi\pi(I=0)|H_W|\overline{K}^0\rangle \equiv A_0 e^{i\delta_0}. \tag{10.4.5}$$

For $A_0 > 0$, where δ_0 is the $I = 0$, $\pi - \pi$ phase shift. In this basis, M_{12} and Γ_{12} would be real if there were no CP violation. Then because CP violation is a small effect, the phases of M_{12} and Γ_{12} are small, and we can usefully work to first order in ϵ. Then (10.4.3) implies

$$\epsilon \simeq \frac{i\,\mathrm{Im}\,M_{12} + \mathrm{Im}\,\Gamma_{12}/2}{\Delta M - i\Gamma/2}. \tag{10.4.6}$$

In the standard basis, we expect Im Γ_{12} to be much smaller than Im M_{12}. This follows because (10.4.5) implies that the contribution to Γ_{12} from the 2π, $I = 0$ states that dominate the decay are real. Thus, on top of the usual suppression of CP–violating effects, Im Γ_{12} should have an additional suppression of at least a few hundred (the ratio of the 2π, $I = 0$ decay rate to everything else). Thus, the phase of ϵ is determined by the phase of the denominator in (10.4.6),

$$\arg(\Delta M - i\Delta\Gamma/2) \simeq 46.2^\circ \tag{10.4.7}$$

By pure coincidence, this is close to 45° because experimentally

$$\Delta M \simeq -\Delta\Gamma/2. \tag{10.4.8}$$

It has become standard, if sometimes confusing, to use these empirical relations to simplify the expression for ϵ,

$$\epsilon \simeq \frac{e^{i\pi/4}}{2\sqrt{2}}\frac{\mathrm{Im}\,M_{12}}{\mathrm{Re}\,M_{12}} \simeq \frac{e^{i\pi/4}\,\mathrm{Im}\,M_{12}}{\sqrt{2}\Delta M}. \tag{10.4.9}$$

One of the striking implications of (10.4.1), the form of the $K_{L,S}$ states, is that a K_L beam should decay into positively and negatively charged leptons at different rates. The decays into positive leptons (antileptons actually) such as $K_L^0 \to \pi^- e^+ \nu_e$, come only from the K^0 component, while the decays into negative leptons, such as $K_L^0 \to \pi^+ e^- \overline{\nu}_e$ come only from the $\overline{K}^0$ component. Thus, there is an asymmetry

$$\delta = \frac{\Gamma(+) - \Gamma(-)}{\Gamma(+) + \Gamma(-)} = \frac{|1+\epsilon|^2 - |1-\epsilon|^2}{|1+\epsilon|^2 + |1-\epsilon|^2} \simeq 2 \text{ Re } \epsilon. \tag{10.4.10}$$

This asymmetry has been observed. The experimental value of δ, $\delta = 0.330 \pm 0.012\%$ gives

$$\text{Re } \epsilon \simeq 1.65 \times 10^{-3}. \tag{10.4.11}$$

We can now use this measurement to get a potentially interesting constraint on the CP–violating phase in the KM matrix.

Using the results of Section 10.2 and 10.3, we find

$$\text{Im } M_{12} \simeq \frac{1}{2m_K} \frac{G_F^2}{16pi^2}$$
$$\text{Im} \left\{ \left[-\xi_c^2 m_c^2 \eta_1 + \xi_t^2 m_t^2 \eta_2 + 2\xi_c \xi_t m_c^2 \ln(m_t^2/m_c^2) \eta_3 \right] \cdot \langle K^0 | O_+^{\overline{d}s} | \overline{K}^0 \rangle \right\}. \tag{10.4.12}$$

In Section 10.2, when we discussed the matrix element of $O_+^{\overline{d}s}$, we assumed that it was real. In the presence of CP violation, as we will discuss in detail in the next section, that is not necessarily true. Thus, we have allowed for the possibility that the matrix element is complex. We have also assumed that the second term in(10.2.2) does not make a significant contribution to IM M_{12} in this basis. This is reasonable because we might expect it to be real in the basis in which Γ_{12} is real, since they are related by analyticity. Γ_{12} and the second term in (10.2.2) are related to the absorptive and dispersive parts, respectively, of the matrix element

$$\int \langle K^0 | T(\mathcal{H}_1(x) \mathcal{H}_1(0)) | \overline{K}^0 \rangle \, d^4x. \tag{10.4.13}$$

Nevertheless, this is not a very strong argument because $\mathcal{H}_1$ involves several different operators with different phases that could contribute differently to the absorptive and dispersive parts. But it is probably the best we can do.

If we give the phase of the matrix element a name,

$$\arg(\langle K^0 | O_+^{\overline{d}s} | \overline{K}^0 \rangle) = 2\xi, \tag{10.4.14}$$

we can combine (10.4.9), (10.4.11), and (10.4.12) with the numerical results found in Section 10.3 to find the constraint

$$10^{-3} \simeq \text{Im } \left\{ e^{2i\xi} \left[-0.7\xi_c^2 + 0.6(m_t^2/m_c^2)\xi_t^2 + 0.4 \ln(m_t^2/m_c^2) \xi_c \xi_t \right] \right\}. \tag{10.4.15}$$

We do not yet know enough about the parameters ξ, the KM angles, and m_t to check (10.4.15). But as more information becomes available, (10.4.15) will become an increasingly important check on this picture of CP violation.

10.5 $K_L \to \pi\pi$ and the Parameter ϵ'

The parameter ϵ describes CP violation in the mixing of K^0 and $\overline{K}^0$, which shows up experimentally because of the for (10.4.1) of the eigenstates of H. One can also imagine that CP violation might show up directly in the weak decays of K^0 and $\overline{K}^0$. Consider, for example, the two–pion final state, which is particularly important because of the $\Delta I = \frac{1}{2}$ enhancement. We have already chosen the phase of the $\Delta I = \frac{1}{2}$ decay amplitude (10.4.5) as part of our definition of ϵ. But in principle, the $\Delta I = \frac{3}{2}$ might have a different phase. Define

$$\begin{aligned} \langle\pi\pi(I=2)|H_W|K^0\rangle &= A_2 e^{i\delta_2} \\ \langle\pi\pi(I=2)|H_W|\overline{K}^0\rangle &= A_2^* e^{i\delta_2}, \end{aligned} \tag{10.5.1}$$

where the form follows from CPT invariance and δ_2 is the $I = 2$, $\pi\pi$ phase shift, incorporating the effect of the final–state interactions. If A_2 is complex in the basis in which A_0 is real, then there is CP violation in the decay. This effect is unlikely to be large. Not only is it CP–violating, but it goes away in the limit in which the $\Delta I = \frac{1}{2}$ rule is exact.

To analyze the effect quantitatively in $K \to 2\pi$ decays, form the measurable quantities

$$\begin{aligned} \eta_{+-} &= \frac{\langle\pi^+\pi^-|H_W|K_L\rangle}{\langle\pi^+\pi^-|H_W|K_S\rangle} \\ \eta_{00} &= \frac{\langle\pi^0\pi^0|H_W|K_L\rangle}{\langle\pi^0\pi^0|H_W|K_S\rangle}. \end{aligned} \tag{10.5.2}$$

Both are direct measures of CP violation, since they are proportional to the $K_L \to 2\pi$ decay amplitude. Evaluation (10.5.2) using (10.4.1) and (10.4.5), (10.5.1) and

$$\begin{aligned} \langle\pi^+\pi^-| &= \sqrt{\tfrac{2}{3}}\langle 2\pi(I=0)| + \sqrt{\tfrac{1}{3}}\langle 2\pi(I=2)| \\ \langle\pi 0 + \pi^0| &= \sqrt{\tfrac{2}{3}}\langle 2\pi(I=2)| - \sqrt{\tfrac{1}{3}}\langle 2\pi(I=0)|, \end{aligned} \tag{10.5.3}$$

we find

$$\eta^+ \simeq \epsilon + \epsilon', \quad \eta^{00} \simeq \epsilon - 2\epsilon', \quad \epsilon' = i \,\mathrm{Im}\, A_2 e^{(i\delta_2 - \delta_0)}/\sqrt{2}\, A_0, \tag{10.5.4}$$

where we have neglected terms of order $\epsilon' A_2/A_0$ and smaller. Experiments so far are consistent with ϵ'/ϵ from zero to a few percent, but experiments in progress should either see it or push the bound on ϵ'/ϵ down well below the 1% level. The experimental measurement of the phase of η^{+-} agrees with the prediction(10.4.9) for the phase of ϵ.

The CP violation in the decays measured by ϵ' vanishes trivially in so-called superweak models of CP violation in which the CP violation is associated with an interaction that is of order G_F^2 in strength. Such an interaction would be too small to show up anywhere except in the $\Delta S = 2$ mass mixing, M_{12}. A superweak force can produce a nonzero ϵ, but it cannot produce CP violation in decay amplitudes (because they are of order G_F).

The six–quark $SU(2) \times U(1)$ model is not a superweak model, since CP violation appears in the order G_F interactions. However, for some time it was thought to be very nearly superweak. The point is that in the four–quark interactions induced by W exchange at $\mu \sim M_W$, all phases can be removed from the $\Delta S = 1$ interactions involving only the light quarks (as is done in the usual parametrization of the KM matrix). Then the weak decays are real (up to final–state interactions) in the standard quark–model basis for the hadron states.

Gilman and Wise were the first to realize that this argument might be seriously wrong. While the couplings of $O_{\pm}^{\bar{u}u}$ are real at $\mu \sim M_W$, those of the $\Delta S = 1$ operators $O_{\pm}^{\bar{t}t}$ and $O_{\pm}^{\bar{c}c}$ are, in general, complex. QCD interactions mix $O_{\pm}^{\bar{t}t}$ and $O_{\pm}^{\bar{c}c}$ with the $SU(3)_L$ octet operators (9.4.4)–(9.4.8) that involve only light quarks and gluons. Thus, nontrivial phases can occur in the low–energy $\Delta S = 1$ weak Hamiltonian. Indeed, there is reason to expect their contribution to be important because they are associated with the "Penguin" operator [O_4 in (9.4.7)] whose matrix elements may be large enough to produce much of the observed $\Delta I = \frac{1}{2}$ enhancement.

Explicitly, suppose we work in the quark basis. Then the $K^0 \to 2\pi$ decay amplitudes are real except for the effect of the "Penguin" operator. Assuming that it is O_4 that has the large matrix element and working to first order in CP–violating phases, we can write

$$A_0^Q \simeq A_0 e^{i\xi}, \tag{10.5.5}$$

where

$$\xi \simeq (\text{Im } h_4(\mu)) \langle 2\pi(I=0)|O_4|K^0\rangle^Q / A_0, \tag{10.5.6}$$

and the superscript Q indicates the quark basis. Note that ξ is proportional to $s_2 s_2 \sin\delta$ because it depends on Im $\xi_c = -$Im ξ_t. This is the same small parameter that suppresses all CP–violating effects.

The ξ defined in (10.5.5) and (10.5.6) is the same as that in (10.4.14) because the matrix element $\langle K^0|O_{\pm}^{\bar{d}s}|\overline{K}^0\rangle$ is real in the quark basis, but to transform to the basis in which A_0 is real, we must redefine the states as follows:

$$|K^0\rangle \to e^{-i\xi}|K^0\rangle, \quad |\overline{K}^0\rangle \to e^{-i\xi}|\overline{K}^0\rangle. \tag{10.5.7}$$

This multiplies the matrix element by $e^{2i\xi}$.

The A_2 amplitude is not infested with "Penguins". It comes purely from the $\Delta I = \frac{3}{2}$ part of the weak Hamiltonian that has a real coefficient. Thus, it is real in the quark basis. But then the transformation to the basis in which A_0 is real introduces a factor of $e^{-i\xi}$, thus

$$A_2 = e^{-i\xi}|A_2|, \tag{10.5.8}$$

and for small ξ, the parameter ϵ' is

$$\epsilon' = -\xi e^{i(\delta_2 - \delta_0)}|A_2|/\sqrt{2}\, A_0 \simeq -e^{i\pi/4}\xi/20\sqrt{2}, \tag{10.5.9}$$

where we have put in the experimental values of $|A_2|A_0 \simeq 1/20$ and $\delta_0 - \delta_2 \simeq \pi/4$ (actually it is $53^\circ \pm 6^\circ$).

Eventually, data on ϵ'/ϵ and the KM angles will continue to improve, m_t will be measured, and perhaps we will even be able to calculate some strong–interaction matrix elements. Progress on any of these fronts could transform (10.5.9) and (10.4.15) into crucial tests of the six–quark model of CP violation. Perhaps the most exciting possible outcome of all this would be some evidence that new physics is required to understand CP violation.

Problems

10-1. Derive (10.2.5) and (10.2.6).

10-2. Reproduce ("derive" is too strong a term) (10.2.7). Note that there are four different ways of inserting the vacuum state.

10-3. In the alternative $SU(2) \times U(1)$ theory of Problem 3-4, suppose that

$$U_1 = c_1 u + s_1 c\,, \qquad U_2 = c_1 c - s_1 u$$

$$D_1 = c_2 d + s_2 s\,, \qquad D_2 = c_2 s - s_2 d$$

In tree approximation, there is no flavor-changing neutral-current effect. But what about radiative corrections? Calculate the leading contribution to the $\Delta S = 2$ part of the effective Lagrangian just below M_W. Assume that all the quark masses are small compared to the W mass and keep only the leading terms. You will probably find it easiest to calculate the appropriate Feynman diagrams in 't Hooft-Feynman gauge. Be sure to explain why the diagrams you calculate give the leading contribution.

10-4. Integrate the renormalization group equation for h_{ct}, with β_{ct} given by (10.3.13).

10-5. Derive (10.4.3) and (10.4.4).

10-6. Relate (10.4.13) to Γ_{12} and the second term in (10.2.2).

10-7. Derive (10.5.4).

10-8. Suppose that some very weak interaction at a large scale M_I produces a $\Delta S = 2$, CP-violating interaction of the form

$$\frac{i}{M_I^2} O_+^{\bar{d}s}$$

Estimate how large M_I would have to be in order for this term to produce all the CP-violating effects that have been observed.

10-9. Consider the effect of the ρ parameter (see Section 8.5) of a doublet of very heavy quarks h and l with usual $SU(2) \times U(1)$ properties but with masses m_h, $m_l \gg M_W$. First show that if both quarks are removed from the theory at the same large scale, $\mu \approx m_h, m_l$, the effect on ρ is a matching correction that has the form

$$\Delta\rho = \frac{e\alpha}{16\pi]sin^2\theta M_W^2}\left\{m_h^2 + m_l^2 - \frac{2m_h^2 m_l^2}{m_h^2 - m_l^2}\ln\frac{m_h^2}{m_l^2}\right\}$$

Now consider the case $m_h \gg m_l \gg M_W$, and remove the quarks from the theory one at a time. You should find that the first and the second terms in the bracket are still matching corrections, while the third arises from the form of the β function in the region $m_l < \mu < m_h$. Calculate it using the renormalization group, and show that it reduces to the result above when you turn off QCD.

Bibliography

[Abbott 81] L. F. Abbott, Nucl. Phys. **B185** (1981) 189. See also L. F. Abbot, M.. T. Grisaru and R. K. Schaffer, Nucl. Phys. **B229** (1983) 372.

[Pais] A. Pais, **Inward Bound**.

[Coleman 73] S. Coleman and E. Weinberg, Phys. Rev. **D7**, 1888, 1973. A classic! The effective action, the renormalization group, introduces dimensional transmutation, and more, are all discussed in the inimitable Coleman style.

[Itzykson] C. Itzykson and J.-B. Zuber, **Quantum Field Theory**, McGraw-Hill, 1980. Almost everything is here, but it is often presented without a global point of view.

Appendix A

Review of Dimensional Regularization

In dimensional regularization (DR), the quantum field theory of interest is extended to $4-\epsilon$ dimensions.[1] Because this extension changes the dimensions of the fields, if the couplings are to retain their canonical dimension, appropriate fractional powers of a dimensional parameter, μ, must be introduced into the couplings of the theory. We will discuss the physical significance of this parameter, and the related question of "In what sense is dimensional regularization a sensible regularization scheme?" However, before we get to these interesting questions, we will briefly review the formalism.

A.1 n Dimensional Integration

Begin with the standard calculation of the n-dimensional volume:

$$\begin{aligned}\pi^{n/2} &= \left(\int_{-\infty}^{\infty} e^{-p^2}\, dp\right)^n = \int e^{-p^2}\, d^n p = \Omega(n) \int_0^{\infty} e^{-p^2}\, p^{n-1}\, dp \\ &= \frac{1}{2}\Omega(n) \int_0^{\infty} (p^2)^{\frac{n}{2}-1}\, e^{-p^2}\, d(p^2) = \frac{1}{2}\Omega(n)\,\Gamma\left(\frac{n}{2}\right) \\ &\Rightarrow \Omega(n) = \frac{2\,\pi^{n/2}}{\Gamma(n/2)}\end{aligned} \tag{A.1.1}$$

Now use this to compute the canonical Feynman graph in κ dimensions:

$$\begin{aligned}\int \frac{(p^2)^\beta}{(p^2+A^2)^\alpha}\,\frac{d^\kappa p}{(2\pi)^\kappa} &= \frac{\Omega(\kappa)}{(2\pi)^\kappa}\int_0^{\infty} \frac{p^{\kappa+2\beta}}{(p^2+A^2)^\alpha}\,\frac{dp}{p} \\ &= \frac{\Omega(\kappa)}{(2\pi)^\kappa}\,A^{\kappa+2\beta-2\alpha}\int_0^{\infty} \frac{y^{\kappa+2\beta}}{(1+y^2)^\alpha}\,\frac{dy}{y}\end{aligned} \tag{A.1.2}$$

[1] I will think of ϵ as a negative number. This makes no difference. The dimensionally regualrized integrals are defined by analytic continuation anyway. I simply prefer to think of the $-\epsilon$ dimensions as "extra" dimensions, beyond the usual 4.

Let

$$x = \frac{y^2}{1+y^2}, \quad y^2 = \frac{x}{1-x}, \quad 1+y^2 = \frac{1}{1-x},$$

$$2\ln y = \ln x - \ln(1-x), \tag{A.1.3}$$

$$2\frac{dy}{y} = dx\left(\frac{1}{x} + \frac{1}{1-x}\right) = \frac{dx}{x(1-x)}.$$

Then (A.1.2) becomes

$$\begin{aligned} &= \frac{\Omega(\kappa)}{(2\pi)^\kappa} A^{\kappa+2\beta-2\alpha} \frac{1}{2}\int_0^1 x^{\beta+\kappa/2-1}(1-x)^{\alpha-\beta-\kappa/2-1}\,dx \\ &= \frac{\Omega(\kappa)}{(2\pi)^\kappa} A^{\kappa+2\beta-2\alpha} \frac{1}{2} \frac{\Gamma(\beta+\kappa/2)\,\Gamma(\alpha-\beta-\kappa/2)}{\Gamma(\alpha)}. \end{aligned} \tag{A.1.4}$$

or

$$\begin{aligned} &\int \frac{(k^2)^\beta}{(k^2+A^2)^\alpha} \frac{d^\kappa k}{(2\pi)^\kappa} \\ &= \frac{A^{\kappa+2\beta-2\alpha}}{(4\pi)^{\kappa/2}} \frac{\Gamma(\beta+\kappa/2)\,\Gamma(\alpha-\beta-\kappa/2)}{\Gamma(\kappa/2)\,\Gamma(\alpha)}. \end{aligned} \tag{A.1.5}$$

Finally, taking $\kappa = 4-\epsilon$, and dividing by $\mu^{-\epsilon}$ to get the dimensions right, we have the final result

$$\begin{aligned} &\int \frac{(k^2)^\beta}{(k^2+A^2)^\alpha} \frac{d^{4-\epsilon} k}{(2\pi)^{4-\epsilon}\mu^{-\epsilon}} \\ &= \frac{A^{4+2\beta-2\alpha-\epsilon}}{(4\pi)^{2-\epsilon/2}\mu^{-\epsilon}} \frac{\Gamma(\beta+2-\epsilon/2)\,\Gamma(\alpha-\beta-2+\epsilon/2)}{\Gamma(2-\epsilon/2)\,\Gamma(\alpha)}. \end{aligned} \tag{A.1.6}$$

Often, for example in calculations of anomalous dimensions, we are only interested in the μ dependence of the diagram. In that case, the only relevant ϵ dependence at the one loop level comes from the $1/\mu^{-\epsilon}$ and the $1/\epsilon$ pole in (A.1.6). In this case, we can rewrite (A.1.6) (for nonnegative integer α and β such that $\beta+2 \geq \alpha$) in the simpler form:

$$\begin{aligned} &\int \frac{d^{4-\epsilon} k}{(2\pi)^{4-\epsilon}\mu^{-\epsilon}} \frac{(k^2)^\beta}{(k^2+A^2)^\alpha} \\ &= \frac{A^{4+2\beta-2\alpha}}{(4\pi)^2\mu^{-\epsilon}} \frac{\Gamma(\beta+2)\,\Gamma(\alpha-\beta-2+\epsilon/2)}{\Gamma(\alpha)} + \cdots. \\ &= \frac{A^{4+2\beta-2\alpha}}{(4\pi)^2} \frac{1}{\epsilon\,\mu^{-\epsilon}} \frac{2\,(-1)^{\beta-\alpha}(\beta+1)!}{(\alpha-1)!\,(\beta+2-\alpha)!} + \cdots. \end{aligned} \tag{A.1.7}$$

where $\cdots$ are finite terms independent of μ. This piece has all the information about renormalization in it.

A.2 Regularization

The physical idea of a regularization scheme is that it is a modification of the physics of the theory at short distances that allows us to calculate the quantum corrections. If we modify the physics only at short distances, we expect that all the effects of the regularization can be absorbed into the parameters of the theory. That is how we chose the parameters in the first place. However, it is not obvious that DR is a modification of the physics at short distances. To see to what extent it is, consider a typical Feynman graph in the unregularized theory in Euclidean space:[2]

$$I = \int \frac{1}{(p^2 + A^2)^\alpha} \frac{d^4p}{(2\pi)^4} \, [dx] \, , \tag{A.2.8}$$

where $[dx]$ indicates the integration over Feynman parameter, and A^2 is an function of the external momenta and the particle masses, and of the Feynman parameters. α is some integer. All graphs ultimately reduce to sums of objects of this form.

In DR, these objects are replaced by integrals over $4 - \epsilon$ dimensional momentum space:

$$I_\epsilon = c(\epsilon) \int \frac{1}{(p_\epsilon^2 + p^2 + A^2)^\alpha} \frac{d^{4-\epsilon}p}{\mu^{-\epsilon} (2\pi)^{4-\epsilon}} \, [dx] \, , \tag{A.2.9}$$

where $c(\epsilon)$ is some function that goes to 1 as $\epsilon \to 0$. In (A.2.9), p^2 retains its meaning as the 4 dimensional length. To see the relation between (A.2.9) and (A.2.8), rewrite (A.2.9) as follows

$$c(\epsilon) \int \frac{1}{(p_\epsilon^2 + p^2 + A^2)^\alpha} \frac{d^{-\epsilon}p}{(2\pi\mu)^{-\epsilon}} \frac{d^4p}{(2\pi)^4} \, [dx] \, , \tag{A.2.10}$$

I have explicitly separated out the "extra" $-\epsilon$ dimensions, so that p^2 is the 4 dimensional length of p.

Now use (A.1.5) to do the integral over the $-\epsilon$ extra dimensions (of course this is not the way we would actually calculate the graph – but it will help us to understand what is happening). The result is

$$I_\epsilon = \int \frac{1}{(p^2 + A^2)^\alpha} \, r(\epsilon) \left(\frac{p^2 + A^2}{4\pi\mu^2} \right)^{-\epsilon/2} \frac{d^4p}{(2\pi)^4} \, [dx] \, , \tag{A.2.11}$$

where

$$r(\epsilon) = c(\epsilon) \frac{\Gamma(\alpha + \epsilon/2)}{\Gamma(\alpha)} \, . \tag{A.2.12}$$

The multiplicative factor, $r(\epsilon)$, goes to 1 as $\epsilon \to 0$. The important factor is

$$\rho^{-\epsilon/2} \quad \text{where} \quad \rho = \frac{p^2 + A^2}{4\pi\mu^2} \, . \tag{A.2.13}$$

This factor also goes to 1 as $\epsilon \to 0$, but here the convergence depends on p and A. Because

$$\rho^{-\epsilon/2} = e^{-(\epsilon \ln \rho)/2} \, , \tag{A.2.14}$$

it follows that

$$\rho^{-\epsilon/2} \approx 1 \quad \text{for} \quad |\ln \rho| \ll \frac{1}{\epsilon} \, . \tag{A.2.15}$$

[2]We discuss one loop graphs, for simplicity. The extension to arbitrary loops is trivial.

Equation (A.2.15) is the crucial result. It shows that when we compute quantum corrections using dimensional regularization, we do not change the physics for p (the loop momentum) and A (which involves external momenta and masses) of the order of μ. However, we introduce significant modifications if either p or A is much larger than μ for fixed ϵ, or if they are both much smaller than μ. The second is an important caveat. DR changes the physics in the infrared as well as the ultraviolet region. If all the particles are massive and all momenta spacelike, that causes no problems, because A^2 is bounded away from zero. However, in general, we will need to keep our wits about us and make sure that we understand what is happening in the infrared regime, in order to use dimensional regularization. As long as the infrared region is under control, dimensional regularization should be a perfectly good regularization technique.

Appendix B

Background Field Gauge

With (1b.2.12), the theory is well defined, but the gauge symmetry has been explicitly broken. That is OK, because we know that gauge invariant Green functions are independent of the breaking, however, it makes the analysis of theory more complicated. There is a way of choosing the gauge that defines the theory, but nevertheless preserves the advantages of explicit gauge invariance. This is the so-called "background field gauge" [Abbott 81]. The idea is to include a classical gauge field, A^μ, and build the gauge fixing term as a function of the quantity

$$G_L \equiv \partial^\mu G_\mu - \partial^\mu A_\mu - i[G^\mu, A_\mu]\,. \tag{B.0.1}$$

The L is a reminder that this quantity depends on the longitudinal component of the gauge field. The point is that while G_L transforms inhomogeneously under (1b.2.5), it is covariant under the **simultaneous** transformations of (1b.2.5) and

$$A^\mu \to \Omega A^\mu \Omega^{-1} + i\Omega\partial^\mu\Omega^{-1}\,. \tag{B.0.2}$$

Thus we can build a gauge fixing term, $f(G, A)$, (in fact, a function of G_L) that is invariant under the simultaneous gauge transformations, for example,[1]

$$\exp\left(-i\int \operatorname{tr}\frac{1}{\alpha g^2}\, {G_L}^2\right)\,. \tag{B.0.3}$$

The Fadeev-Popov determinant defined with such a gauge-fixing term, $f(G, A)$, is separately invariant under (1b.2.5) and (B.0.2). The first follows from (1b.2.8) and (1b.2.9). The second follows immediately from the simultaneous invariance of $f(G, A)$. Thus the whole action in (1b.2.12) has the simultaneous gauge invariance.

To see how this symmetry constrains the form of the Green functions of the theory, add a source term,

$$2\operatorname{tr} G^\mu J_\mu\,, \tag{B.0.4}$$

and use (1b.2.12) to construct the generating functional, $W(J, A)$, that now depends on both the source and on the classical gauge field. There is also dependence on sources for matter fields, but we will not write this explicitly. The symmetry properties can be seen more readily if we rewrite the source term as

$$2\operatorname{tr}(G^\mu - A^\mu)J_\mu + 2\operatorname{tr} A^\mu J_\mu\,. \tag{B.0.5}$$

[1] from here on, we assume the normalization, (1.1.14), unless explicitly stated to the contrary

The first term in (B.0.5) is invariant under (1b.2.5), (B.0.2), and

$$J^\mu \to \Omega J^\mu \Omega^{-1} . \tag{B.0.6}$$

The second term, when exponentiated, simply gives a constant factor that can be taken outside the functional integral. Thus we can write

$$W(J,A) = \tilde{W}(J,A) + 2\,\mathrm{tr}\, A^\mu J_\mu , \tag{B.0.7}$$

where $\tilde{W}(J,A)$ is invariant under (1b.2.5) and (B.0.6).

When we construct the generating functional for 1PI graphs, the effective action, we get

$$\Gamma(\mathcal{G},A) = W(J,A) - 2\,\mathrm{tr}\,\mathcal{G}^\mu J_\mu = \tilde{W}(J,A) - 2\,\mathrm{tr}\,(\mathcal{G}^\mu - A^\mu) J_\mu , \tag{B.0.8}$$

where $\mathcal{G}$ is the classical field corresponding to the quantum gauge field, G, defined by

$$\mathcal{G} = \frac{\delta W}{\delta J} = \frac{\delta \tilde{W}}{\delta J} + A . \tag{B.0.9}$$

These two relations show that $\Gamma(\mathcal{G},A)$ is invariant under the simultaneous gauge transformations, (B.0.2) and

$$\mathcal{G}^\mu \to \Omega \mathcal{G}^\mu \Omega^{-1} + i\Omega \partial^\mu \Omega^{-1} . \tag{B.0.10}$$

Now even though the Γ function has the gauge symmetry, it is still not very useful, because with two gauge fields, $\mathcal{G}$ and A, we can build many more invariant terms than with one alone. However, we have not yet used the fact that the classical gauge field, A, is arbitrary, because it is part of the gauge fixing. In particular, we can choose

$$A^\mu = \mathcal{G}^\mu , \tag{B.0.11}$$

and use

$$\Gamma_{BF}(\mathcal{G}) \equiv \Gamma(\mathcal{G},\mathcal{G}) , \tag{B.0.12}$$

which is invariant under (1b.2.5) alone.

Next, let us show that we can use the object, $\Gamma_{BF}(\mathcal{G})$, to calculate the gauge invariant physical quantities of interest. We begin by including in our Lagrangian a set of sources, J, for gauge invariant operators. These operators are constructed out of the quantum gauge field, G, and any matter fields that are relevant. The advantage of including these sources is that we can then look only at the vacuum energy in the presence of the sources, and by differentiation with respect to the sources, reconstruct the matrix elements of gauge invariant operators. Any physical quantity can be related to suitable matrix elements of gauge invariant operators, thus the vacuum energy in the presence of arbitrary gauge invariant sources contains all the information we need to do physics.

The quantization of the theory proceeds in the same way in the presence of the sources. The only difference is that now the Γ functions we construct depend on the sources, J. Now the object

$$\Gamma(G,A,J) \tag{B.0.13}$$

has the property of independence of A at an extremum with respect to G because at such an extremum, it a vacuum energy. Thus for example, if

$$\Gamma_1(G_0(A,J),A,J) = 0 \tag{B.0.14}$$

then $\Gamma(G_0(A,J),A,J)$ is independent of A. Thus

$$\frac{d}{dA}\Gamma(G_0(A,J),A,J) = 0 \tag{B.0.15}$$

$$\frac{\partial G_0}{\partial A}\Gamma_1(G_0(A,J),A,J) + \Gamma_2(G_0(A,J),A,J) \tag{B.0.16}$$

and therefore

$$\Rightarrow \Gamma_2(G_0(A,J),A,J) = 0\,. \tag{B.0.17}$$

Now in particular, we can choose

$$A = g_0(J) \tag{B.0.18}$$

which is a solution to the equation

$$g_0(J) = G_0(g_0(J),J)\,. \tag{B.0.19}$$

Then

$$\Gamma(G_0(A,J),A,J) = \Gamma(g_0(J),g_0(J),J) = \Gamma_{BF}(g_0(J),J) \tag{B.0.20}$$

and

$$\frac{\partial}{\partial A}\Gamma_{BF}(A,J)|_{A=g_0(J)} \tag{B.0.21}$$

$$= \Gamma_1(g_0(J),g_0(J),J) + \Gamma_2(g_0(J),g_0(J),J) = 0\,. \tag{B.0.22}$$

So the vacuum energy in the presence of gauge invariant sources is given by Γ_{BF} at an extremum. Thus we can use ΓBF to compute arbitrary vacuum matrix elements of gauge invariant operators, from which we can construct any physical quantity.

In fact, the way the background field gauge is used in practice, is to compute anomalous dimensions of gauge invariant operators. In such calculations, it is convenient to deal with objects that are not gauge invariant, such as matrix elements of the gauge invariant operators with gauge fields or matter fields. It is often easiest to find the renormalization group equations by studying these gauge variant objects. But the anomalous dimensions of the gauge invariant operators obtained by this procedure must be gauge invariant in background field gauge because **they could have been obtained by studying only the matrix elements of gauge invariant operators from Γ_{BF}.**

The only problem with Γ_{BF}, as defined in (B.0.12), is that it is unnecessarily difficult to calculate. We must evaluate diagrams with both external gauge field lines and with insertions of the classical gauge field, A. We can calculate Γ_{BF} much more easily by changing variables in the original Lagrangian, and taking the quantum field to be

$$\tilde{G}^\mu \equiv G^\mu - A^\mu\,. \tag{B.0.23}$$

In the effective action, the classical field gets translated in the same way,[2]

$$\tilde{\mathcal{G}}^\mu \equiv \mathcal{G}^\mu - A^\mu\,, \tag{B.0.24}$$

and the effective action in the translated theory is

$$\tilde{\Gamma}(\tilde{\mathcal{G}},A) = \Gamma(\tilde{\mathcal{G}}+A,A)\,. \tag{B.0.25}$$

Now clearly, we can calculate Γ_{BF} as

$$\Gamma_{BF}(A) \equiv \Gamma(A,A) = \tilde{\Gamma}(0,A)\,, \tag{B.0.26}$$

[2]This is one of the great advantages of the effective action in the study of symmetry breaking – see [Coleman 73].

This is a great improvement. Now we must calculate 1PI diagrams with insertions of the classical gauge field, but we do not need to have any external quantum gauge field lines! We will use this form. Our definitions have been chosen to agree with those of [Abbott 81], except that we have normalized the fields so that the gauge transformations do not involve the gauge coupling, so that the coupling appears in the coefficient of the kinetic energy term, (1b.2.2) and (B.0.3), and the ghost kinetic energy term in the Fadeev-Popov determinant. The Feynman rules are shown below, with gauge fields indicated by double lines and ghost fields by dotted lines. The external gauge field is denoted by a double line ending in Ⓐ. We have not shown any vertices with a single quantum gauge field, because these do not contribute to the 1PI diagrams that we must calculate to evaluate Γ_{BF}.

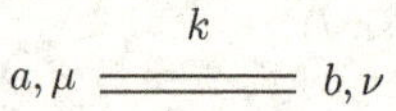

$$\frac{-i\,g^2\,\delta_{ab}}{k^2+i\epsilon}\left[g_{\mu\nu}-\frac{k_\mu k_\nu}{k^2}(1-\alpha)\right] \tag{B.0.27}$$

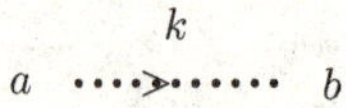

$$\frac{i\,g^2\,\delta_{ab}}{k^2+i\epsilon} \tag{B.0.28}$$

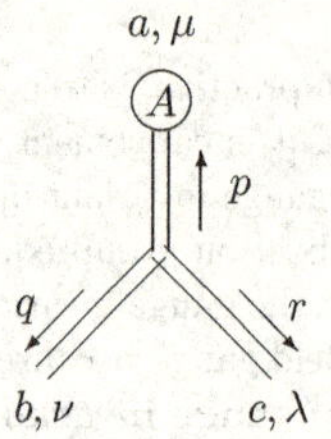

$$\frac{1}{g^2}\,f_{abc}\,[g_{\mu\lambda}(p-r-q/\alpha)_\nu$$
$$+g_{\nu\lambda}(r-q)_\mu$$
$$+g_{\mu\nu}(q-p+r/\alpha)_\lambda] \tag{B.0.29}$$

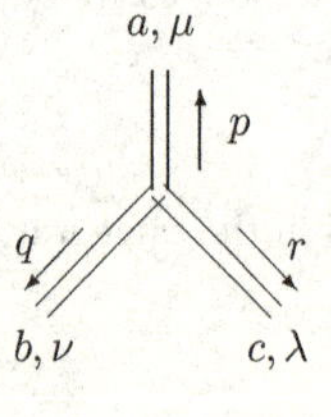

$$\frac{1}{g^2}\,f_{abc}\,[g_{\mu\lambda}(p-r)_\nu$$
$$+g_{\nu\lambda}(r-q)_\mu$$
$$+g_{\mu\nu}(q-p)_\lambda] \tag{B.0.30}$$

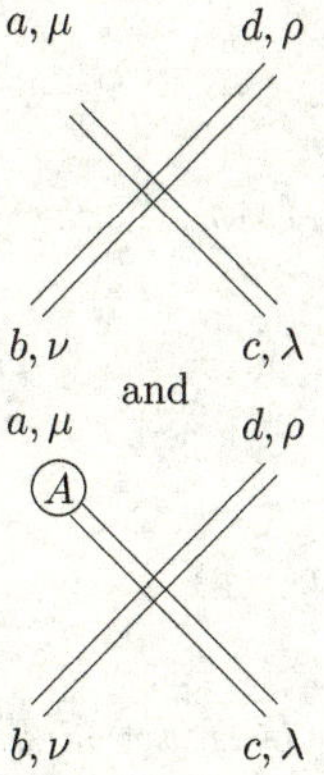

$$\frac{-i}{g^2}[f_{abx}f_{xcd}(g_{\mu\lambda}g_{\nu\rho}-g_{\mu\rho}g_{\nu\lambda}) + f_{adx}f_{xbc}(g_{\mu\nu}g_{\lambda\rho}-g_{\mu\lambda}g_{\nu\rho}) + f_{acx}f_{xbd}(g_{\mu\nu}g_{\lambda\rho}-g_{\mu\rho}g_{\nu\lambda})] \tag{B.0.31}$$

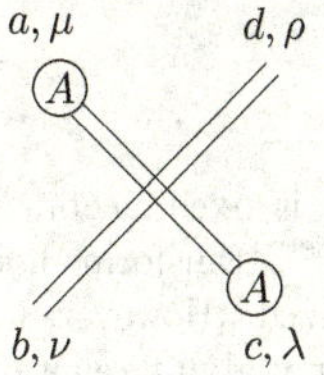

$$\frac{-i}{g^2}[f_{abx}f_{xcd} \cdot (g_{\mu\lambda}g_{\nu\rho}-g_{\mu\rho}g_{\nu\lambda}+\tfrac{1}{\alpha}g_{\mu\nu}g_{\lambda\rho}) + f_{adx}f_{xbc}(g_{\mu\nu}g_{\lambda\rho}-g_{\mu\lambda}g_{\nu\rho}-\tfrac{1}{\alpha}g_{\mu\rho}g_{\nu\lambda}) + f_{acx}f_{xbd}(g_{\mu\nu}g_{\lambda\rho}-g_{\mu\rho}g_{\nu\lambda})] \tag{B.0.32}$$

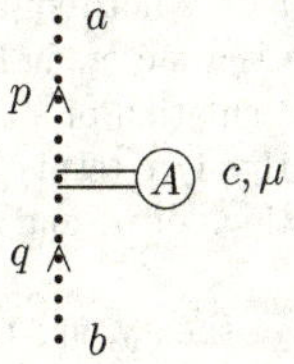

$$-\frac{1}{g^2}\, f_{abc}\,(p+q)_\mu \tag{B.0.33}$$

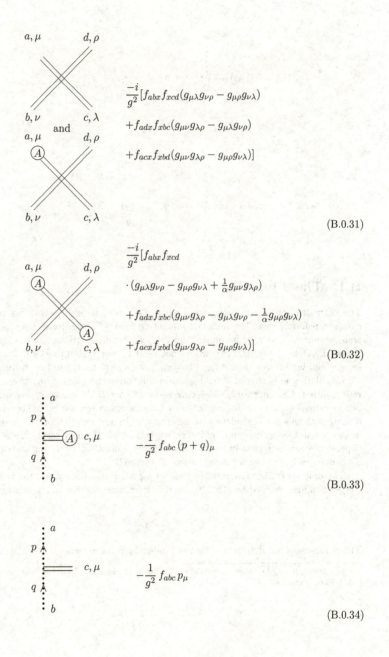

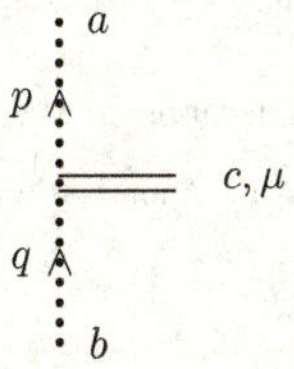

$$-\frac{1}{g^2}\, f_{abc}\, p_\mu \tag{B.0.34}$$

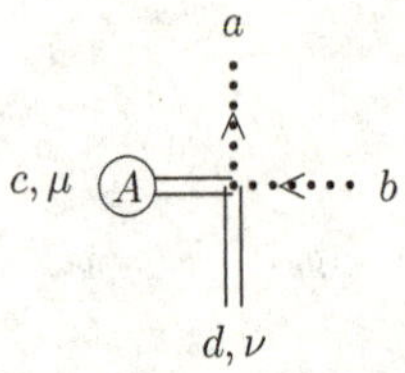

$$\frac{-i}{g^2} f_{acx} f_{xdb} g_{\mu\nu} \tag{B.0.35}$$

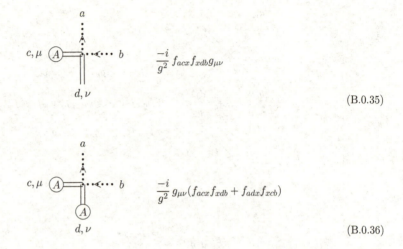

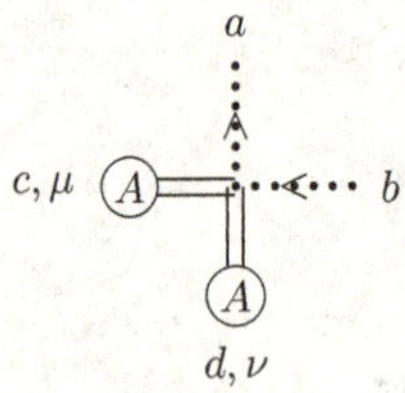

$$\frac{-i}{g^2} g_{\mu\nu} (f_{acx} f_{xdb} + f_{adx} f_{xcb}) \tag{B.0.36}$$

B.1 The β function

The normalization of the background gauge field, A^μ, is fixed, in our notation, by the form of gauge transformations. The translated quantum field, $\tilde{G}^\mu$, transforms homogeneously under gauge transformations and may be multiplicatively renormalized. However, in calculating the effective action, $\Gamma_{BF}(A)$, using (B.0.26), we do not need to worry about renormalization of the quantum gauge field, because we do not need to couple sources to it. It can be left unrenormalized.

Calculating in background field gauge still involves the gauge fixing parameter α, which is also renormalized. Landau gauge, which corresponds to $\alpha = 0$, does not requires such a renormalization, but we cannot go to Landau gauge directly because of the $1/\alpha$ factors in Feynman rules. One way of dealing with this is to calculate with arbitrary α. In a calculation of any gauge invariant quantity, in particular the background field effective action, all the $1/\alpha$ terms cancel, because the gauge dependence can appear only in the form of renormalizations of α, and then α can be set to zero. Thus one can ignore the renormalization of α as well.[3]

The only remaining renormalizations, in a pure gauge theory, are the renormalizations of the gauge couplings, defined by (1b.2.2). At tree level, the effective action looks just like (1b.2.2),

$$-\sum_a \frac{1}{4g_a^2} A_{a\mu\nu} A_a^{\mu\nu} . \tag{B.1.1}$$

The β functions are determined by the infinite loop corrections to the effective action which can

[3]In practice, it may be easier to calculate in Feynman gauge, as in [Abbott 81], and include counterterms for α, but it is not necessary.

renormalize (B.1.1). At one loop, these are completely determined by the two point diagrams,

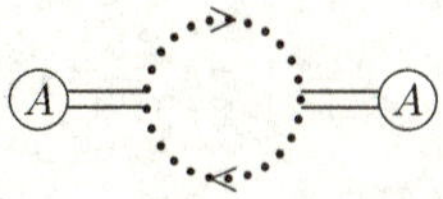

(B.1.2)

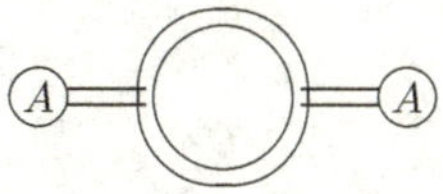

(B.1.3)

Note that the fact that only 2-point functions are required to find β is a consequence of the explicit gauge invariance of the effective action that guarantees that the infinite contributions will have the form of (B.1.1). This makes the calculation of β in background field gauge much simpler than a calculation in a standard covariant gauge.

Another way of saying this is that because the explicit gauge invariance fixes the normalization of A^μ, the renormalization group equation satisfied by the effective action is

$$\left(\mu\frac{\partial}{\partial\mu}+\beta_g\frac{\partial}{\partial g}\right)\Gamma_{BF}(A)=0\,. \tag{B.1.4}$$

Since (B.1.4) does not involve derivatives with respect to A, we can calculate β by looking at the any n-point function, in particular the 2-point function. Including the one loop corrections from (B.1.2) and (B.1.3), we have

$$\Gamma^{\mu\nu}_{2ab}(p)=\delta_{ab}\left[\frac{1}{g_a^2}+b\,\ln\left(\frac{-p^2}{\mu^2}\right)+\text{finite terms}\right]\left[g^{\mu\nu}p^2-p^\mu p^\nu\right]\,, \tag{B.1.5}$$

where

$$b=\frac{11}{3}\frac{C_a}{16\pi^2}\,, \tag{B.1.6}$$

with

$$C_a\,\delta_{ab}=\sum_{de}f_{ade}f_{bde}\,. \tag{B.1.7}$$